Texts in Computer Science

Series Editors

David Gries, Department of Computer Science, Cornell University, Ithaca, NY, USA

Orit Hazzan , Faculty of Education in Technology and Science, Technion—Israel Institute of Technology, Haifa, Israel

More information about this series at http://www.springer.com/series/3191

Richard Hill • Stuart Berry

Guide to Industrial Analytics

Solving Data Science Problems
for Manufacturing and the Internet
of Things

 Springer

Richard Hill 🆔
Department of Computer Science
University of Huddersfield
Huddersfield, UK

Stuart Berry
Department of Computing
and Mathematics
University of Derby
Derby, UK

ISSN 1868-0941   ISSN 1868-095X   (electronic)
Texts in Computer Science
ISBN 978-3-030-79106-3   ISBN 978-3-030-79104-9   (eBook)
https://doi.org/10.1007/978-3-030-79104-9

This Springer imprint is published by the registered company Springer Nature Switzerland AG
The registered company address is: Gewerbestrasse 11, 6330 Cham, Switzerland

To Megan and Daniel.

Foreword

Industrialisation is an essential part of global economic development. Central to this development is technology; creating of new ways of doing things to enhance quality, repeatability and to discover new frontiers of value creation.

The automotive and aerospace industries have been prime movers in the advancement of the application of technology to improve lives and generate wealth. Both the outputs of these industries, and the environments in which they are produced, require a continual application and adaptation of technology to make our activities safer, more affordable and ultimately more sustainable, as the global community becomes aware of our collective need to reduce the consumption of increasingly scarce natural resources.

Thus, innovation is one of our primary tools to address current and future challenges. Innovation gives the ability to respond rapidly to emergent situations and to make reasoned sense of historical experience, so that we can learn from the past to inform the future.

Data has always been central to industry; without measurements, quantities, reporting and accounting, we would not have been able to make the advancements that have been witnessed through industrialisation.

However, it is more recent developments in computing technologies that are creating new ways to use data to create even more value and more advanced products.

Through widespread application of wireless sensor networks, embedded systems, cloud computing and ubiquitous high speed network infrastructure, we can identify hidden patterns in operational data, store and process vast quantities of data and constantly refine computational algorithms to search, categorise and predict new behaviours in a complex, inter-connected world.

This use of technology to collect, organise, process and consume data provides industry with the ability to monitor performance, automate decision making through condition monitoring and predictive maintenance, create seamless supply chain linkages through the close integration of industrial processes and logistics, leads us to discover and release value streams that were not seen prior to the adoption of analytics technologies. These technologies are key as we enter into a more model-based engineering (MBSE) approach to industrial innovation.

The 2017 UK Government Industrial Strategy[1] has been a catalyst of technology awareness. While the 'Tier 1' leaders of industry have understood that technology is central to their competitive advantage, other suppliers further down the supply chain have not always been able to keep pace with the early adopters.

Financial constraints such as limited investment have restricted the extent by which small and medium sized enterprises (SME) can explore the benefits of technological innovation until the technology itself becomes more affordable.

We are now at the beginning of an exciting era where technology is relatively inexpensive and the key differentiator between a business that is agile and fit for the future, and one that may struggle to remain sustainable, is the knowledge of how to use data—specifically the techniques of analytics—to maintain their competitive edge.

April 2021

Dr. Paul Needham, Ph.D. CEng FIET
Visiting Professor
University of Huddersfield
Huddersfield, UK

---

[1] https://www.gov.uk/government/publications/industrial-strategy-building-a-britain-fit-for-the-future.

Preface

Overview and Goals

Technology is a key enabler of business, and as computation and storage costs become lower, what was once a vision of *computing as a utility* is now becoming the reality. Cloud computing models have illustrated how new business value and competitive advantage can be created from new ways of collaboration; inexpensive microprocessors and pervasive broadband networks are facilitating processing power that can be embedded into a constant stream of new applications.

As we start to see the possibilities of physical objects, that are inter-connected to share data, we can start to contemplate the potential of what an Internet of Things (IoT) environment might look like. From an industrial perspective, and especially manufacturing, there is the need to process and move physical objects to create business value.

As organisations strive to differentiate themselves from their competition, new ideas to increase sales revenue places hitherto unrecognisable demands upon the whole manufacturing supply chain.

What were once complex, but manageable challenges in planning, scheduling, production control and logistics, are rapidly becoming situations that are impossible to supervise without automation.

How we automate our industrial processes, to some extent assumes that we know what needs to be automated and that we have the necessary experience and skills to be able to bring the automation to fruition in a reliable way.

At the heart of any investigation into industrial operations is an understanding of:

1. what data is required;
2. what data is available;
3. what data processing needs to take place;
4. how to communicate the results of the analysis to a business audience.

Guide to Industrial Analytics: Making sense of data science for manufacturing and the Industrial Internet of Things is an attempt to address the need of organisations who can see the possibilities of an inter-connected industrial world but do not know how to make effective use of their data. It is commonplace for software vendors to sell 'black box' solutions that only solve one specific problem, yet many industrial challenges can be solved with some knowledge of specific techniques that are commonly utilised in the field of *data science.*

Often, guides to data science target audiences that are fluent in abstract mathematics. This serves little purpose for busy professionals who need to concentrate on their business needs.

This book has deliberately focused on the need to understand the practical application of data science techniques to solve industrial challenges, with minimal knowledge of mathematics required. Where the mathematics is essential, a detailed explanation is provided.

As such, the key objectives for this book include:

- to present an understanding of the fundamental approaches to analysing data that is commonly found in industrial environments;
- to understand the procedures and thinking around the selection and cleaning of industrial data;
- to demonstrate how we can apply different aspects of data science to discover interesting insight within data, using commonly available tools;
- to explore ways in which we can use existing data to make predictions about the future;
- to explore the ways in which visualisation can be used to enable the improved comprehension of industrial data;
- to understand the application of simple techniques to common situations, while also being aware of their limitations;
- to identify areas of further study in what is a fast-moving domain.

Target Audiences

The use of data to obtain new value and create opportunities for industrial businesses has a broad appeal. We have deliberately focused on delivering a book that shows how to apply data science techniques to industrial scenarios, and therefore the text is couched as a set of learn-by-doing exercises.

We have also taken a pragmatic stance in terms of the tools used to illustrate the examples. All software used is either freely available (open source) or is generally regarded as pervasive; it is likely that industrial organisations will have access to spreadsheets such as Microsoft Excel, or alternatives, for instance.

As such, *business leaders*, *industrial managers* and *supervisors* will find the combination of *just enough* mathematics and extensive practical explanations of value to them. Many traditional texts are long on theory and short on application. The know-how in this book will help them make more informed operational decisions, which in turn will improve the quality of data available for strategic planning.

Application developers who work on industrial enterprise IT systems will also be able to observe the type of analysis that industrial personnel want to do, and it is hoped that this book will inform the design and specification of updates and modifications to such systems infuture.

University instructors will find that this book is a suitably concise volume that can help get advanced undergraduate and postgraduate students applying data science techniques quickly. Many new university courses are including data science, or aspects of it as part of the curricula, but it is the real-world application of these techniques that is often lacking.

Finally, *technical consultants* and commercially oriented *researchers*, who work directly with industry to deliver tangible improvements, will find the collection of how-to articles for common scenarios of use to them in their business, especially the section on visualisation, in order to successfully communicate insight and conclusions to their clients.

Organisation and Suggested Use

This book is organised into three parts:

- Part I introduces the concepts of manufacturing analytics and data science.
- Part II describes a range of techniques and approaches to solving problems.
- Part III illustrates the application of methods and processes by way of industrial examples.

Guide to Industrial Analytics should be used as a comprehensive introduction to the use of data science techniques in real-world situations. Part 1 of the book provides the general foundation of the important concepts and is a good place to start for readers new to the topics.

Since our presentation of topics is rooted firmly in practicality, we recommend that Part I is reviewed by all readers. While there are countless texts on the specifics of data analytics, our presentation of the application of these techniques is relatively unique and there is much to be learned by reading about a topic or concept int he context of the industrial environment.

Part II should be seen as a walkthrough in the application of analytics techniques that have been proven to work. While many more exotic techniques exist, we have focused on approaches where there is the most return for the effort expended. These approaches have wide applicability and will significantly enhance the reader's ability to conduct useful and profitable industrial analytics.

Part III looks at the wider context of industrial acceptance of analytics, demonstrating that the barriers to successful adoption are not always limited by technical prowess.

A series of appendices present essential technical material to support the bulk of the text.

The book is designed to help readers acquaint themselves with practical techniques for dealing with industrial data, before becoming a reference text when the important processes are practised and understood.

For *university instructors*, we suggest the following program of study for a 12-week semester delivery pattern:

- Weeks 1–2: Part 1;
- Weeks 3–7: Part II;
- Weeks 8–11: Part III;
- Week 12: Assessment.

Part I explores the context of industrial data, how it is used and what we can achieve with it. It provides a practical definition of data science and explores how we might apply techniques to different situations. There is also an introduction to the tools that can be used to perform the analysis.

Part II is a more in-depth look at a range of techniques that we use to find insight from data. Using a tutorial-based approach, there are specific examples that demonstrate data science skills in practice. There are also exercises for the readers to complete and reinforce their own knowledge.

Part III demonstrates the application of approaches and techniques to real life. These examples help illustrate how we can combine different techniques to solve a particular industrial query.

Learning Activities

Each chapter concludes with a set of review questions and learning activities that make specific reference to the material in each chapter. There is also an additional set of more open questions that will require further investigation from the reader. Such questions help embed the material that has been learned, so that it can be applied to a number of different situations. These questions will be useful to university instructors who can set them as homework activities outside of class.

Hands-on Exercises

Much of the application of this work requires proficiency in the use of tools. Much of the *mystique* of data analytics is knowing what tool works in what set of circumstances and we have deliberately focused upon the use of tools that are commonly available in the industrial environment. Exercises are used throughout the book to illustrate not only the 'what' but also the 'how' of data analytics. Our intention is for readers to develop sufficient skills to use the techniques as tools when they are faced with an industrial analytics job in the future, hence the strong thread of learning-by-doing.

Acknowledgements

The authors would like to express their gratitude to the many organisations and business owners from the Midlands and Yorkshire regions of the UK, for their cooperation and engagement with research and consulting projects over the past 25 years. Much of the insight within this book has come from practical experience, experimentation, evaluation and the general fieldwork of being involved in industrial operations.

In addition, the chapters explaining data analysis and statistics have been developed over 12 years of teaching undergraduate students, with particular contributions from Richard Self and Dr. Pritesh Mistry.

Huddersfield, UK   Richard Hill
Derby, UK   Stuart Berry
April 2021

Contents

Part I Introductory Concepts

1 An Introduction to Industrial Analytics 3
  1.1 What Is Analytics? 3
  1.2 Breaking Boundaries 4
      1.2.1 The Industrial Internet of Things 6
      1.2.2 Disruption Means Change 8
  1.3 Industry 4.0 9
  1.4 Opportunities for Smart Businesses 11
  1.5 What Is Data Science? 12
  1.6 Why Do We Need Data Science? 13
  1.7 A Process for Data Science 14
      1.7.1 Data Preparation 15
      1.7.2 Data Exploration 17
      1.7.3 Model Selection 17
      1.7.4 Evaluation 18
  1.8 Do We Need Machine Learning for Industrial Analytics? 19
  1.9 Learning Activities 19
  References 20

2 Data, Analysis and Statistics 21
  2.1 Introduction 21
  2.2 The Need for Analysis and Statistics 21
  2.3 Qualitative and Quantitative Data 22
  2.4 Data Terminology 23
  2.5 Data Quality 23
  2.6 Scales of Measurement 25
      2.6.1 Nominal Data 25
      2.6.2 Ordinal Data 28
      2.6.3 Interval Data 28
      2.6.4 Ratio Data 28

2.7   Central Tendency   29
        2.7.1   Mean..   29
        2.7.2   Median   30
        2.7.3   Mode..   30
2.8   Dispersion   31
        2.8.1   Range   32
        2.8.2   Interquartile Range   32
        2.8.3   Variance   32
        2.8.4   Standard Deviation   33
        2.8.5   Frequency   33
2.9   Histogram   35
        2.9.1   Cumulative Frequency Graph   36
2.10   Shape of the Data..................................   36
        2.10.1 Normal Distribution   37
        2.10.2 Uniform Distribution   42
        2.10.3 Bimodal Distribution   42
        2.10.4 Skewed Distributions   43
2.11   Visualising Data   44
        2.11.1 Pie Charts   45
        2.11.2 Bar Charts   47
        2.11.3 Line Charts   49
        2.11.4 Scatter Plots................................   50
2.12   Learning Activities   50
References   51

3   Measuring Operations.......................................   53
    3.1   Introduction   53
    3.2   Using Assumptions................................   54
    3.3   Operations Concepts   55
            3.3.1   Cycle Time   55
            3.3.2   Lead Time   57
            3.3.3   Takt Time   57
    3.4   Using Concepts to Understand Systems   57
    3.5   Resource Utilisation   59
    3.6   Learning Activities   62
    References   63

4   Data for Production Planning and Control   65
    4.1   Historical Attitudes Towards the Use of Data   65
    4.2   Need for Data Within the Production Area.................   66
    4.3   Planning Problems Resulting from Lack of Appropriate
            Data..   67
    4.4   Planning with Appropriate Data........................   68
    4.5   Need for Optimality in Production Control and Scheduling   69

4.6   Deriving Generic Models for Planning and Control.   75
4.7   Production Planning in Manufacturing: Small Case Study
       Results   78
4.8   Planning and Control in the Case Study Firms   79
4.9   Manufacturing Production Systems in Case Study Firms   80
4.10   Summary   81
4.11   Learning Activities   82
References   83

Part II   Methods

5   Simulating Industrial Processes   87
5.1   Understanding Business Operations   87
5.2   Queues and Queueing   88
5.3   Modelling an Industrial Process   91
5.4   Designing a Process Simulation   92
5.5   Building the Simulation in Ciw   93
5.6   Confidence   96
5.7   Conclusion   98
5.8   Learning Activities   99

6   From Process to System Simulation   101
6.1   Simulating Industrial Systems   101
6.2   Example: Joinery Manufacturer   102
6.3   Building the Simulation.   103
6.4   Managing Resource Utilisation   111
6.5   Product Mixes   113
       6.5.1   Sash Windows   114
6.6   Conclusion   124
6.7   Learning Activities   124

7   Constructing Machine Learning Models for Prediction   127
7.1   Introduction   127
7.2   Data and Prediction.   128
       7.2.1   Example 1: Job Time Prediction Under Varying
                Demand   129
7.3   Assessing the Predictive Power of a Model   133
       7.3.1   Root Mean Squared Error (RMSE)   133
       7.3.2   Mean Absolute Error (MAE).   134
       7.3.3   Mean Absolute Percentage Error (MAPE)   134
       7.3.4   Coefficient of Determination (R^2)   134
       7.3.5   Underfitting and Overfitting   134

	7.3.6	Cross-Validation	135
	7.3.7	Learning Curves	136
	7.3.8	Validation Curves	136
7.4	How to Improve Model Accuracy		137
	7.4.1	Feature Selection	137
	7.4.2	Example 2: Improving the Model with Additional Information (Multiple Regression)	138
	7.4.3	More Data	139
	7.4.4	Compare Models	140
	7.4.5	Example 3: Multiple Job Types (Model Comparison)	140
7.5	Generating Data Via Simulations		141
	7.5.1	Example 4: Simulating Data Under Uncertainty	142
	7.5.2	Kernel Density Estimation and Sampling	145
7.6	Worked Examples in R		145
	7.6.1	Linear Regression Model	145
	7.6.2	Multiple Regression Model	148
	7.6.3	Cross-Validation	149
	7.6.4	KDE Estimation of Distribution	150
References			151

Part III   Application

8	**Case Study: Confectionery Production**		155
8.1	Introduction		155
	8.1.1	Company Organisation	156
	8.1.2	Production	156
8.2	Hard Boiled Confectionery Organisation and Planning		158
	8.2.1	Unit Operation	159
	8.2.2	Scheduling	160
	8.2.3	Model to Determine Optimal Long Term, Monthly Production Plans	160
	8.2.4	Implementation of Monthly Planning	164
	8.2.5	Allocating Production Pairs	166
	8.2.6	Implementation of Daily Pair Selection	169
8.3	Impact of Lack of Information on Company Profits		172
8.4	Conclusion		173
8.5	Learning Activities		174
	8.5.1	Machinery and Staffing	176
	8.5.2	Sales Data	177

9	**Minimum Information Set for Effective Control**		183
9.1	Information Flows Within an Organisation		183
9.2	Deriving Minimum Information Requirements		186

9.3 Order Book-Based Systems 186
  9.3.1 Deriving an Order Book (OB) Forecasting Model
        Where Orders for New Jobs Arrive Randomly
        in Time 189
  9.3.2 Variability in the Final Product Introduced
        at All Stages 190
  9.3.3 Simple Order Book System.................... 191
  9.3.4 Enhanced Order Book Models 192
  9.3.5 Forecast Accuracy/Validation 194
  9.3.6 Extending the Investigation by Including Data
        from All Stages 195
  9.3.7 Variability Added (only) at the Final Stage 196
  9.3.8 Conclusion: Order Book (OB)-Based Approaches
        to Forecasting 198
9.4 Work Book (WB) Systems 199
9.5 Evaluating WB and OB When Stage Productions Have
    Been Balanced 204
  9.5.1 Conclusions and Recommendations MIR 205
9.6 Data Requirements for Planning and Control 208
9.7 Minimal Information in Flow Shops with CONWIP
    Control...................................... 208
  9.7.1 Flow Shops with More Than One Like Processor
        at Each Stage 210
  9.7.2 Job Shops 211
  9.7.3 Effect of Growth on Planning and Control
        in a Flow Shop 212
  9.7.4 Results from Three-Stage Models 212
  9.7.5 Results from 10-Stage Models 213
9.8 Conclusion 214
9.9 Learning Activities 215

10 Business Adoption of Analytics 217
10.1 Introduction 217
  10.1.1 Intelligent Manufacturing 217
  10.1.2 Compounded Challenges for SMEs 218
  10.1.3 Regional Challenge 218
10.2 A Model of Engagement........................... 219
  10.2.1 Proving the Return on Investment 220
  10.2.2 Digital Enablers Network (DEN)................. 220
10.3 University Capability 222
  10.3.1 Case Study: DEN in Action 222

10.4   Discussion   223
       10.4.1   Benefits to SMEs   224
       10.4.2   Benefits to den Members   224
       10.4.3   Benefits to Academia   224
       10.4.4   Human Factors   225
10.5   Conclusions   225
10.6   Future Work   226
       References   227

Appendix A: Statistics   229

Appendix B: Simulation Library—Ciw   243

Appendix C: Production Planning Programs in MS Excel   245

Appendix D: Scheduling Jobs Through a Workshop   263

Index   273

Contributors

Stuart Berry University of Derby, Derby, UK

James Devitt University of Huddersfield, Huddersfield, UK

Richard Hill University of Huddersfield, Huddersfield, UK

Sam O'Neill University of Derby, Derby, UK

Abbreviations

3DM	Data-driven decision-making
ACID	Atomicity, consistency, isolation and durability
AD	Automatic differentiation
AE	Autoencoder
ANN	Artificial neural network
AOSD	Aspect-oriented software development
API	Application programming interface
AQL	Annotation query language
ARD	Automatic relevance determination
ASR	Automatic speech recognition
BDA	Big data analytics
BNN	Binary neural net
BPTS	Back-propagation through structure
BPTT	Back-propagation through time
CART	Classification and regression trees
CCA	Canonical correlational analysis
CEP	Complex event processing
CNN	Convolutional neural network
COCO	Common objects in context
COTS	Commodity off-the-shelf
CPPN	Compositional pattern-producing network
CQL	Cassandra query language, cyber-query language, common/contextual query language
CRF	Conditional random field
CTC	Connectionist temporal classification
CV	Cross-validation
DAD	Discover, access, distil
DAG	Directed acyclic graph
DBN	Deep belief network
DCGAN	Deep convolutional generative adversarial networks
DHSL	Distributed Hadoop storage layer
DNN	Deep neural network
DT	Decision tree
EBM	Energy-based model

ECL	Enterprise control language
EDA	Exploratory data analysis, event-driven architecture
ELU	Exponential linear unit
EPN	Event processing nodes
ERF	Extremely random forest
ESP	Enforced subpopulations
FUSE	Filesystem in userspace
GA	Genetic algorithm
GAN	Generative adversarial network
GBM	Gradient boosting machine
GEOFF	Graph serialisation format
GMM	Gaussian mixture model
GRU	Gated recurrent unit
HAM	Hierarchical attentive memory
HAR	Hadoop archive
HMM	Hidden Markov model
HPCC	High performance computing cluster
HPIL	Hadoop physical infrastructure layer
HTM	Hierarchical temporal memory
IDA	Initial data analysis
IIoT	Industrial Internet of Things
IoT	Internet of Things
JAQL	JSON query language
JSON	JavaScript object notation
KFS	Kosmos file system
KNN	k-nearest neighbours
KPCA	Kernel principal component analysis
KSVM	Kernel support vector machine
LOOCV	Leave one out cross-validation
LReLU	Leaky ReLU
LSTM	Long short-term memory
LTU	Linear threshold unit
LZO	Lempel-Ziv-Oberhumer
MAE	Mean absolute error
MAPE	Mean absolute percentage error
MCMC	Markov chain Monte Carlo
MDM	Master data management
MDP	Markov decision processes
ML	Machine learning
MLP	Multi-layer perceptrons
MSA	Microservices architecture
NB	Naive Bayes
NEAT	Neuro-evolution of augmenting topologies
NLP	Natural language processing
NN	Neural network

NTM	Neural turing machine
OCR	Optical character recognition
OLAP	Online analytical processing
OLS	Ordinary least squares regression
OLTP	Online transactional processing
PAC-MDP	Probably approximately correct in Markov decision processes
PCA	Principal component analysis
PMML	Predictive Model Markup Language
PReLU	Parameterised ReLU
RBM	Restricted Boltzmann machines
RDD	Resilient distributed database
ReLU	Rectified linear unit
ResNet	Residual neural network
RF	Random forest
RL	Reinforcement learning
RMSE	Root mean squared error
RNN	Recurrent neural network
RNTN	Recursive neural tensor network
RTRL	Real-time recurrent learning
RVM	Relevance vector machine
S4	Simple scalable streaming system
SANE	Symbiotic adaptive neuro-evolution
SD	Standard deviation
SGD	Stochastic gradient descent
SIFT	Scale-invariant feature transform
SOA	Service-oriented architecture
SRN	Simple recurrent network
SVD	Singular value decomposition
SVM	Support vector machine
TDA	Topological data analysis
TF	TensorFlow
TFIDF	Term frequency inverse document frequency
UDAF	User-defined aggregate function
UDTF	User-defined table-generating function
UIMA	Unstructured information management architecture
VC	Vapnik Chervonekis dimension
VLAD	Vector of locally aggregated descriptors
W3C	World Wide Web Consortium
WFST	Weighted finite-state transducers
XML	Extensible Markup Language
YARN	Yet another resource manager

Part I
Introductory Concepts

Part I introduces important concepts upon which the remainder of the book is founded.

An Introduction to Industrial Analytics

1

1.1 What Is Analytics?

There was a time, not so long ago, when the topic of analytics caused confusion. 'Don't you mean data *analysis*?' was the likely retort. Data and latterly business analysis are subjects that industry are comfortable with. We analyse data to work out what happened or to understand our businesses better. We perform business analysis when we want to explore the potential of a new idea or to evaluate the performance of a business function.

The act of analysis is to break a topic down into manageable parts so that we can start to make sense of a complex topic. It is a scientific discipline that uses methods and techniques to provide rigour, repeatability and an auditable path to an explanation. As such, analysis uses historical data to aid comprehension of the past.

Analytics is the process of taking the results of the analysis and then using this to create tools that can predict future behaviour. Forecasting is another mature subject, where mathematical models are formulated specifically to make predictions.

Continued advances in technology have made computational resources and data storage both accessible and inexpensive. Cloud computing provided the foundation for big data to proliferate, which in turn has made data more available than ever.

This combination of technology, accessibility, free and low-cost toolsets, together with mature subjects such as statistical analysis, networked data access and forecasting tools, has made it feasible to apply traditional techniques at a larger scale.

As such, the software vendors have brought analysis and prediction together to form *analytics*; its distinction with *analysis* is of decreasing relevance today, as most discussions use analytics and analysis interchangeably.

Used specifically in the business domain, analytics is a means of determining the profitability and revenue opportunities of an organisation, as well as the potential shareholder return for larger institutions. The outputs from analytics processes are used to keep the business competitive, which is typically an activity that requires agility in fast-moving markets.

© Springer Nature Switzerland AG 2021
R. Hill and S. Berry, *Guide to Industrial Analytics*, Texts in Computer Science,
https://doi.org/10.1007/978-3-030-79104-9_1

Analytics is typically described as comprising four separate stages, of which there is a logical progression from one stage to the next:

1. *Descriptive analytics*—the use of data to understand the past. '*What has happened?*'
2. *Diagnostic analytics*—discovering patterns in the data to identify root causes. '*Why did it happen?*'
3. *Predictive analytics*—analysing historical performance to predict future performance. '*What will happen?*'
4. *Prescriptive analytics*—using optimisation techniques to inform the direction of action. '*How can we make it happen?*'

The ability to have analytics automated so that the actions of an organisation are prescribed is a compelling vision to aspire to, and it is understandable why businesses are keen to develop the necessary capabilities.

Each of the stages assumes that the previous capability is in place and operational. Diagnostic analytics cannot function unless descriptive analytics is available to present the data properly. Similarly, predictive analytics requires insight such as root causes to be incorporated into prediction models.

A primary challenge that prevents realisation of prescriptive analytics is not whether an organisation has insufficient data, or indeed so much data that it is difficult to understand which data is of most value. It is a general lack of capability to establish a robust foundation of descriptive analytics capability that can withstand the weight of enquiry from subsequent analytics tiers.

With all of the talk of the potential for big data, many organisations come to the conclusion that they do not have enough data to embark upon an analytics journey. However, it is common for businesses not to be aware of the data they possess, nor of the rate that they are producing it.

Much of industrial data is inaccessible, insofar as the business not having the skills or awareness to use the data for different purposes.

The data required for analytics activities can be found in the first instance by reviewing any metrics that the organisation uses for performance appraisal. These measures include things that can be counted—referred to as *discrete* items—such as whether something is completed or not and metrics that are measured on a continuum such as delivery time, weight and price.

Databases of transactional data are useful places to find data for analytics enquiries. Spreadsheets are another source of data, as well as hardware and plant that may be using sensors to collect real-time process data.

1.2 Breaking Boundaries

In our modern age of 'connected-ness' we can access data from outside the enterprise, such as weather, financial or social data for example. This data can then be combined or *fused* with internal data to increase the possibilities for innovative analytics to take place.

Industry 4.0 is a movement that brings together networked data, automation, analytics and the wholesale digitalisation of industrial operations. Sharing data beyond the traditional boundaries of an organisation means that whole supply chains or industry sectors can benefit from greater access to data [1].

This is another reason for industrial enterprises to adopt analytics capabilities, especially if competitors have already started to do so. Depending on the industry there may be a very short period between sharing data for competitive advantage and integrating processes for business survival.

However, while the Internet is a widely understood platform where information can be shared freely and rapidly, industrial use of the Internet is more limited. Organisations have demonstrated a reluctance to adopt Internet technologies for industrial applications [3], citing the potential challenges of losing valuable intellectual property (IP) [2] through security breaches, as well as being uncertain as to how Internet technologies might change the fabric of an organisation, or indeed how they might disrupt the fundamental business models.

There is considerable evidence of localised information sharing within industrial organisations, where machine-to-machine (M2M) communications—the practice of networking industrial plant to exchange real-time operational data—but this is usually limited to a constrained set of plant that has the technological capability. As such, if data gathering is only part of the activities of an industrial unit, there is a limit as to the value that can be derived from this beyond the individual process that is monitored.

This situation presents two challenges for industrial organisations. First, there is the investment in technology to enable information sharing. This might be in the form of new plant, which then has to justify its return on investment (ROI) purely based on M2M capabilities. Alternatively, it could be investment through the retrofitting of M2M technology onto existing plant. Recent advances in technology have seen costs dropping dramatically, and this makes the case for retro-fitting technology much stronger. This is a difficult situation though, as the tradition of ROI requires an investment to justify itself by improving the specific application that is to be acquired. But, the significant potential benefits of M2M can only be realised once a critical mass of change has occurred, which is generally beyond the scope of an individual industrial process.

The second challenge is perhaps more of an obstacle for organisations. If we assume that a sufficient proportion of the company's operations are monitored with the assistance of M2M technologies, what is the justification for then including an external network—the public Internet—as part of the data sharing ecosystem?

Companies that use M2M to collect operational data already have more descriptive data to mine for new insight. Surely, the risks to the organisation's business model are increased if that valuable data is shared outside the walls of the factory.

1.2.1   The Industrial Internet of Things

What differentiates the Industrial Internet of Things (IIoT) [4] from traditional industrial M2M technology and infrastructure is the ability to embrace computational
power and storage that is beyond the size and capability that any one industrial
organisation could justify (and afford).

M2M technologies permit processes to be monitored, controlled and managed,
providing a detailed stream of data that then becomes an input into higher-level
operational decisions. If a sensor is installed above a conveyor belt in a production
line and the sensor can detect the presence of a part and produce a date and timestamp
for an electronic record, we can deduce the following:

- the time and date that a particular part was on the conveyor belt;
- whether there are parts being moved or not (which might be downtime due to a
  fault);
- the time between parts, which enables the calculation and reporting of output rate;
- depending on the sensor type or its location, we can identify different parts.

All of the above can be collated from just one sensor, and we have not connected that
sensor to another piece of plant. Using the data to create a record in database means
that we can collect information over a period of time and then look for patterns within
the data. This helps us understand the behaviours of an individual process, but it does
rely on human beings to review the data and make sense of it, in the context of other
data sources. For example, what is the connection between production output rate
and the mix of orders?

Very quickly, the ability to pose questions about operations becomes overshad-
owed by the need to find the data to arrive at a rigorous conclusion that we can
derive value from. We can arrive at some conclusion as to why a process is running
slowly. We shall find it more challenging to understand why the whole factory of
many pieces of plant tends to run slower from time to time.

If we are to tackle the bigger questions of operational efficiencies for an industrial
unit, we need to be able to (a) gather the important data and (b) understand the
questions that need to be asked to drive the appropriate analysis.

These capabilities require both sufficient storage to collect the data, together with
the computational power and analysis methods to discover the insight. Such capabilities have traditionally been the preserve of very large organisations who can
afford the upfront investment in IT infrastructure. The advent of the cloud computing model, where computing as a utility has become available to the masses, has thus
facilitated an uptake, albeit slowly, of the remote storage and analysis of operational
data, otherwise referred to as *big data analytics* (BDA).

Whereas M2M technology permits a process to be monitored, it does not provide
the capability to archive the data for subsequent analysis. The analysis may take
different forms (use different algorithms) and consume different input data (different
sources of data) depending on the questions being asked. This scenario requires

infrastructure that most companies cannot justify financially were they to acquire it for themselves.

Thus, the IIoT brings together the power and flexibility of cloud computing, big data analytics and M2M technologies, to create an environment where a holistic view of an organisation can be formed.

When the data for a collection of related processes is collated and combined with other transactional data such as order/product mixes, economic data, weather forecasts and organisational strategies, previously unanswerable queries can be explored. The inclusion of M2M is a key component part of the system, not only for the sensing of conditions, but also because of the potential for the control of a process.

Condition monitoring is a subject of interest to organisations who use fixed or mobile plant as part of their business. Sensors can detect various operating parameters of equipment and relay this to a cloud-based database. BDA approaches are then used to review the operational data, to see if it matches the data from components that have failed previously. The data for previous failures can then be used to identify the set of conditions prior to failure and then signal to the business that a component is close to failure and thus requires attention. For an organisation that uses a lot of plant, such an approach can be a less costly alternative to traditional *planned maintenance* where parts are routinely replaced to prevent costly downtime, irrespective of whether the part is approaching failure or not.

We can now extend this scenario a little further so that plant starts to become self-aware, as it receives tailored reporting back from the cloud-based BDA enabling self-diagnostics to be performed, *local to the process*. This predictive capability is perhaps one of the most tangible arguments at the moment for organisations to comprehend and is commonly used as the means to justify expenditure on IIoT equipment.

Undoubtedly, one of the main drivers for interest in IIoT is that of comprehension. Without M2M technology we cannot comprehend the detailed inner-workings of an organisation. Production supervisors and managers have wrestled with this challenge for decades and have devised approaches to operations management such as optimised production technology and lean, for example, that attempt to address the need to manage holistically, in the absence of detailed process data. However, the adoption of M2M technology in itself does not necessarily guarantee instant success.

Collecting vast quantities of data from operations does not tackle the underlying issue of comprehension; once the data has been collected, what should be done with it now?

IIoT is the conglomeration of technologies that facilitates the collection, storage and processing of operational data that was previously uneconomical to perform. As the cost of technology continues to plummet and more industries realise the benefits of large-scale holistic business analysis and prediction, the case for adopting IIoT becomes even stronger.

1.2.2   Disruption Means Change

The reasonable man adapts himself to the world: the unreasonable one persists in trying
to adapt the world to himself. Therefore all progress depends on the unreasonable man.
George Bernard Shaw.

Change is often touted as a significant barrier for organisations to accommodate.
The introduction of new ways of doing things can manifest resistance from staff
who appear to be comfortable with the existing status quo. Experienced managers
realise that it is not so much the concept of change that people rail against, it is the
prospect of *disruptive* change. Most people are settled with change if they expect it—
developing a career, starting a family, acquiring skills, etc.,—but it is the imposition
(or perception) of disruption of change that can stimulate an organisational backlash.

Adoption of IIoT technologies presents an interesting blend of skills and knowledge attributes for organisations. Since the technology is pervasive, providing a new
set of lenses with which to scrutinise business operations, it requires involvement
from many stakeholders. Some of the skills, such as retro-fitting technology to plant,
or extending network infrastructure, can be acquired through established training
programs. However, the elicitation of new insight and synthesis of this new insight
into existing business operations can be challenging.

Many industries have mature reporting functions that produce outputs for managers to sift through. An IIoT-rich environment provides predictions based on analysis
that has been performed on bigger than ever datasets. How might staff react to data
that suggests a solution to a problem that appears counter-intuitive? How might staff
react to the need to fundamentally rearrange business functions on a regular basis,
to give the agility required to operate more efficiently?

IIoT adoption places more emphasis upon organisational knowledge. The data
that is collected and the analysis that is automated can lead to challenging questions
for those who live by the rules of the accounting balance sheet. It used to be that
the executive tier of organisations was often accused of lacking data literacy; the
situation is rapidly changing in that there are more knowledge-based demands being
made of management personnel, with the practice of 'war gaming', where different
business scenarios are explored, is becoming even more fruitful as the underlying
data is more detailed and trustworthy than ever before.

The paranoia surrounding IIoT is somewhat justified. Our adoption of technology
is mostly driven by the benefits it offers rather than being tempered by the risks that
it might introduce and so there are sufficient public breaches of information security
to make organisations wary of increasing the reach of their networks.

In a lot of cases though, the use of cloud computing infrastructure is a significant upgrade for information security. The pay-per-use model of computing as a
utility means that small organisations enjoy the benefits of enterprise-level security.
Those remote repositories will be secured and maintained at a much greater level of
scrutiny than if the data was stored locally within an organisation. As such, the real
vulnerabilities lie in the same place that they always have done with the user.

If anything, the prospect of IIoT should be seen as a prompt to review existing information practices within an industrial environment. A wireless link to plant may be easier to install, but the data needs to be transmitted securely. If there is a Bring Your Own Device (BYOD) approach for staff, additional measures to protect these potential vulnerabilities need to be appraised and implemented so that the IP is genuinely secure which all leads to a more general awareness of information security being essential for all of the staff in an organisation being necessary prior to embracing IIoT.

There are also situations where a process is fundamental to the operation of a business, and any disruption to it would cause the business harm. In this scenario it is wise to automate the monitoring, but not the control until the organisation is satisfied that anything—person or system—that interfaces with the process is fully prepared for the implications of change.

All of this talk of introducing change implies that an organisation can develop the capability to accommodate change, but also that it can sustain change. The strategic aspirations of an organisation must recognise the potential effects of increased agility, on both processes and people, and the fact that IIoT facilitates and enables the transformation of businesses into being more knowledge-centric. New ways of designing products digitally enable approaches such as additive manufacturing (*3D printing*) to realise manufactured goods predominantly from information only. As with any shift in thinking, these data-centric technologies will enable new business models to flourish at the expense of those who cannot adapt sufficiently rapidly.

1.3 Industry 4.0

The Fourth Industrial Revolution (Industry 4.0) is a vision of globally connected business that shares data, intelligence and control to enhance, optimise and create sustainable industrial practices. In the same way that steam-powered mechanisation and automation stimulated the First Industrial Revolution, the second revolution came from electrification, followed by the introduction of computing technology as the basis of the Third Industrial Revolution. Where Industry 4.0 differs from Industry 3.0 is that it is an attempt to merge the physical world with the connected, digital world. As described earlier, industry already makes use of technology to connect some functional aspects together, but Industry 4.0 develops this further by way of *cyber-physical systems* (CPS).

A CPS do not replace a human with a robot just yet; it is a concept that enables humans to be a connected, integral part of an autonomous or semi-autonomous system. A current example is that of a 'cooperative robot' or *Cobot*. In a car assembly plant, humans are required to take parts and assemble them together with other parts. Some of these parts are difficult to lift as they are beyond the capability of most workers. Cobots can be used to lift and position heavy or cumbersome items, ready for a human to make the final adjustments and fixings. For some time robots have been used in manufacturing to replace human workers where repetitive or dangerous tasks take place. But in these situations it is also hazardous for a human to be in the presence of the robot as the robot lacks self-awareness. Cobots are designed specifically to work alongside a human worker, including removing itself if it might cause the worker harm.

Central to the vision of Industry 4.0 is the ability for a system to be *smart*. Smart supply chains are composed of smart factories, smart vehicles, smart machines and even smart products. Smartness indicates that data is exchanged and acted upon after some reasoning has taken place. From the explanation of IIoT technologies earlier, the advent of cloud computing and BDA means that we do not need a device to be able to reason itself, to get the operational benefits of a system that reasons. The process of reasoning, informed by BDA, can be carried out in a cloud based on data received from myriad sensor endpoints at the edge of networks.

This capability enables organisational and process memories to be developed that go beyond querying an enterprise resource planning (ERP) system that holds transactional records. It imbues systems with knowledge of their own history, and it encodes the valuable knowledge that constitutes an organisation's intellectual property. If we look at different organisations within a particular industrial sector, they will mostly be using similar resources and methods. However, there will be differences in the way that they use their resources, which is the fundamental basis of any differences in intellectual property, and ultimately their potential to create value.

As technology advances so does choice. Consumers are becoming more demanding, and there are more opportunities for value creation as industries serve ever-more specialised needs and desires from their markets. This differentiation requires more business and industrial agility, such as the ability to produce 100 product lines instead of 10. This increased variation affects an organisation's ability to comprehend the changes required as a company without IIoT cannot perform the analysis to guide management decisions fast enough.

For example, in a factory there may exist more than one route through the plant to manufacture an item. Production supervisors route items in an attempt to maximise the efficiency of the plant operations. However, beyond a certain point it becomes impossible to select an optimum route in real time, which is exactly the scenario where IIoT can assist.

Industry 4.0 thinking assumes a capability to support decision-making, by abstracting management away from detail through intelligent and self-aware systems . It does this by emphasising the integration of the two categories of systems within organisations. The first is *operational technology* (OT), which is the systems that are used to monitor, control, report and manage the 'value-added processes' in the *value chain*. Such technologies are the plant controls and the M2M and associated networking infrastructure.

The second category of system is *information technology* (IT), which is the transactional systems that most business users interact with for functions such as sales order processing, invoicing, human resource management and planning and finance. Bringing together OT and IT facilitates new levels of control by at least providing more data upon which decisions can be taken. This must be supplemented by scalable

BDA resources, which more likely than not will reside remotely in cloud computing systems.

One aspect of Industry 4.0 that is potentially disruptive is its ability to integrate organisations at the level of process rather than product. Industries, particularly manufacturing, have evolved into enterprises that make discrete items or offer discrete services to other manufacturers. Industry 4.0 implies an even tighter integration of organisations which may drastically increase the portfolio of suppliers and customers that a business transacts with. For example, a light engineering job shop might have operated in a way where its machines operate at a low level of utilisation so that a wide variety of order types could be accommodated at short notice. But what if this manufacturer could take electronic orders for a product that only requires some time on one of its machines?

The ability to 'sell' manufacturing hours in a digital market is something that is made possible by an Industry 4.0 environment. Whereas we may be used to large companies undertaking a range of processes, it is feasible that the same processes could be automatically devolved to a virtual collection of smart enterprises, who together operate as if they were one homogenous factory of old.

1.4 Opportunities for Smart Businesses

From the discussion so far we can see that there are are numerous opportunities for industrial organisations to make incremental improvements to their operations through the introduction of IIoT technology. These changes are an essential first step along the road to Industry 4.0 thinking. Indeed, the rate at which an organisation can adapt itself may be ultimately what determines whether the organisation is sustainable in the long term.

Fundamental to Industry 4.0 is the ability to be able to monitor and continuously improve productivity. The increase in productivity can then be used to create larger profit margins and also to create the cashflow to reinvest back into process improvements. If the cost benefits are passed to customers then there is the potential to improve the competitive position of the organisation. Industry 4.0 thinking has the potential to provide greater opportunities for those enterprises who can do a job well; the digitalisation of market places opens up business to new customers and de-emphasises the traditional power of marketing and branding.

Customer service is central to a lot of industrial organisations, whether it be a business-to-consumer (B2C) or business-to-business (B2B) relationship. Industry 4.0's holistic view can only support the enhancement of a customer-centric view of operations. Connected enterprises will develop a perspective (informed by their data) that can lead to improvements that will benefit all of the stakeholders concerned. These benefits are likely to be derived from an ability to *optimise* operations across the supply chain.

At present this occurs mostly with tiered industries—the outputs of Tier 2 manufacturers are orchestrated by the size of their Tier 1 customer. For all of the organisations outside of this arrangement, optimisation, at best, will occur at process level and in all likelihood will stop at the boundary between the organisation and an external supplier/customer. IIoT technology makes agile, real-time decisions to be made and acted upon in a much more agile way, and the integration of the supply chain will increase efficiencies for all parties, including the end-user.

As the shift in demand for skills translates into job opportunities, there will be premiums for early entrants into a knowledge-/data-centric employment market. Increased demand for information security knowledge, data science, cloud computing and BDA skills will create new opportunities for on-demand training services who will be attempting to service the medium-term skills gaps in industry.

Central to both IIoT and Industry 4.0 is the generation, consumption and analysis of data. The range of skills and knowledge required for the handling of data in this new age of industrial development is quite broad, and much of it is considered to be under the umbrella term of *Data Science*.

1.5 What Is Data Science?

Like many computing trends, data science has attracted a lot of attention. Employers are reporting data science 'skills shortages', there are training programs promising your conversion into a *Data Scientist*, and there is a good deal of mystique around what a data scientist actually does, if they actually exist.

If we scrutinise the variety of explanations of data science and we look at the skills that are being taught, it is clear that a common understanding of data science is yet to be realised. For instance, some courses teach statistics, which is a mature topic in its own right. Similarly, other courses illustrate statistical enquiry using a programming language or application. Python and R are examples of this.

At the beginning of this chapter we looked at the difference between data analysis and data analytics; in perhaps the same way that not much is new about data analytics, data science is a collection of techniques, approaches and methods that are very well established. In fact, what is new about data science boils down to two developments:

1. Artificial intelligence (AI) techniques have made the transition from laboratory research to application domain as a result of cheap access to computing technology via the cloud. Cloud computing is therefore an emergent theme that is making a lot of established technology accessible for more general use;
2. As a result of (1), we are now interested in applying established mathematical and statistical models to real-world problem domains, and this has stimulated interest in the rigour of the methods we use. Thus, there is broader applicability of the thinking that underpins the scientific method.

This interest in the methods associated with data science has thus opened the field up to a new generation of individual, and this is now gaining traction as organisations are starting to reap the benefits of using mathematical approaches to evaluating their business operations. There is also an associated increase in research effort, resulting in new refinements and enhancements to statistical and AI algorithms of the past.

For the purposes of this text, we shall use the following definition:

> Data science is the transformation of data using the scientific method and mathematics into valuable insights, decisions and actions.

The definition is quite inclusive, and this is deliberate. Data science should be an inclusive discipline, and a lot of data analysis can be performed with the straightforward application of simple tools. To be consistent with the approach of this book we shall use simple methods that provide the greatest return for the least effort.

1.6 Why Do We Need Data Science?

Business operations are complex, and we invest time and energy into gaining experience, as well as learning approaches to dealing with the complexity via *abstraction*. In a factory floor a supervisor might use checklists to keep track of important events, and plant operators might use monitoring tools such as statistical process control (SPC). While these approaches help manage individual processes, and we assume that measuring the detail will help monitor and control the outcomes of the whole system, in reality this is not the case.

First of all, it is rare that every process is monitored to the same extent. It would probably produce too much data to manage, and the collection of the data might interfere with the actual process of creating a product.

Second, monitoring individual processes does not always make the system variation visible to the people who are charged with the responsibility of coordinating the activities.

Third, even if variation is detected it is not always feasible to re-calculate a schedule rapidly enough, meaning that there is a significant lag between detection and action.

Data science helps us deal with these circumstances. The technologies associated with data science enable us to store and categorise data automatically as processes execute, meaning that we can concentrate on the management of the process itself.

Second, data science can enable us to use historical data to identify patterns. If a process deviates from 'normal operations' a difference in the pattern would be identified and can either be brought to the attention of the process supervisor or in a more advanced environment can be a trigger for a controlled action to take place, as in the case of a CPS.

Finally, data science approaches and techniques, through automation, can facilitate the reassessment of a situation based upon the data that the system is producing; it is not possible for humans to perform analysis with such rigour in as short a timeframe, especially when faced with the multitude of potential data sources from an entire organisation. When combined with trigger-based system responses, an industrial system that is controlled by data science-based approaches can offer enhanced performance (usually through optimisation) over that from a human supervisor.

In summary, data science can

- Produce insightful data in short timescales;
- Use historical data to expose trends in operational data;
- Use trends in data to identify exceptions and to provide a series of different management options;
- Combine historical operational data with forecasted business data, to derive predictions.

The transformation of industry is such that data science is moving from being a desirable function that assists process managers in their roles, to becoming an essential organisational capability that managers need to deal with operational complexity.

1.7 A Process for Data Science

As alluded to in the working definition earlier it is useful to have a process that we can follow when conducting scientific work with data. The process helps us to ensure that we take a consistent approach and that we can have confidence in the results we obtain. Since we have a need to automate our data science approach—to keep up with the data that is being generated—we want to ensure that what is being generated or analysed is correct.

There are six stages as follows:

1. *Define the goal of the study*—This is an important first stage as we should not be embarking upon open-ended enquiry. We should take the time to formulate a question that needs to be answered. As the data science capability develops a library of operational questions will be created as each new study is initiated, and these can be repeated (and automated) to improve the capabilities of operations intelligence;
2. *Data retrieval*—Data retrieval at the commencement of data science thinking is different to collecting data in a mature analytics environment. The questions posed by stage one may initiate some additional data gathering capability such as a sensor. Alternatively, the enquiry might be satisfied by existing sources of data.
3. *Data preparation*—Any data that has been retrieved directly from plant via an OT system may require conditioning before it is usable as part of a data science study. This transformation might be range from the application of signal processing techniques through to reformatting some data fields or adding a date and timestamp. As technology advances smaller devices have greater computational power. The use of *edge* devices might mean that you can perform some of the data preparation local to the source of the data.

4. *Data exploration*—This is the stage where there is perhaps one of the more explicit demands for someone with data science experience. The data is scrutinised and evaluated to investigate whether any patterns or correlations can be visualised. This stage makes the subsequent *model selection* stage much easier, though some data scientists conflate both stages together into one activity.

5. *Model selection*—Once the data has been prepared and some initial exploration has occurred, we are now ready to investigate the data by way of some mathematical algorithms. This stage may find that one algorithm is useful, or it might require a combination of algorithms to produce a set of results that meet our initial criteria for confidence.

6. *Presenting the findings*—The final stage of a data science project is to present the results for communication. The method of presentation is dependent upon your audience. Illustrating the outcomes to executive board members will be different to showing the data to fellow data science colleagues. However, the results that you find might enable a deeper enquiry to be made, and therefore you will be automating the current query so that it can be embedded into your organisation's data science capability. In this case you will need to ensure that the data is presented correctly for the system to accept and categorise along with its other data.

Depending on the starting point for your investigation you may spend more or less time on each of the stages. The path through the stages is not always linear; you might find during the data exploration stage that you need more data—either a greater volume of the existing data types, or—new data types to supplement the existing dataset. You would then return to the data retrieval stage to solicit what you need.

Stages three, four and five are perhaps considered the most technical and specific to the skills of a data scientist. The first is the data preparation stage.

1.7.1 Data Preparation

While there is a range of approaches that can be used to discover new insight from data we have to make sure that we are using data that is representative of reality. It is also important that the data is prepared in a way that facilitates analysis. It should be formatted so that it can be analysed and perhaps processed further in an organised way.

Table 1.1 illustrates how we might format some data. If you are familiar with database design you will recognise some of the ways that we describe data.

Every row in the table is a collection of *attributes*, and together they constitute a *record*. Data scientists usually refer to each row as a *data point*. Each part of the data point might be referred to as a *feature* or *dimension*.

To facilitate subsequent data processing, the type of each attribute should be specified. The primary types are as follows:

Table 1.1 List of transactions from an organisational database. Rows of data represent *data points*. Columns of data represent *attributes/features/dimensions* (terms used interchangeably)

RecordID	WorkCentre	StartDate	OrderID	Completed	UnitCost	Rework
1	Cut	10-01-2021	10056	10-01-2021	14.35	N
2	Cut	10-01-2021	10073	10-01-2021	22.41	Y
3	Drill	09-01-2021	10021	10-01-2021	8.76	N
4	Edge	09-01-2021	10013	11-01-2021	2.56	N
5	Drill	09-01-2021	10034	11-01-2021	9.65	Y
6	Assembly		10056	11-01-2021	22.88	N
7	Paint		10073	11-01-2021	2.74	N
8	Assembly	10-01-2021	10073	11-01-2021	37.82	N

1. *Binary*—A binary attribute indicates that it can be only one of two values. The *Rework* column illustrates that a job can be either performed for the first time or repeated (rework) if there is a defect to be repaired.
2. *Categorical*—Where there are more than two potential values, and the values can be specified as falling into a particular group, we identify the data by a label that represents the grouping. The *WorkCentre* column identifies the area of production where the processing has taken place.
3. *Integer*—Integers are used when whole numbers are to be recorded. In this case, *OrderID* is an attribute of type integer.
4. *Continuous*—Non-integer data which includes a decimal part is represented in the *UnitCost* column.

Contrary to what some software vendors might have you believe, it is not best practice to 'throw all of the data at the algorithm'. Including data that is not relevant to the query being posed may distort the output. Experience guides a data scientist as to which attributes to include, but for more complex investigations it may be necessary to return to this stage to include or exclude attributes to observe the effect on the eventual outcome.

The process of selecting attributes to include is often referred to as *feature engineering*. For instance, if we were investigating workcentres that do not produce *Rework*, we might start looking at the assembly and paint workstations first.

For some reason there is no data recorded for the *StartDate* in data points 6 and 7. This can be a frequent occurrence with IIoT data, where sensors may not be reliable or might become damaged. The absence of data in itself could be interesting as in this case it might point to a system fault. However, for subsequent analysis to take place we may need to deal with a `null` entry.

The easiest approach is to remove the data point, though this is only possible when there is sufficient remaining data on with which to complete the investigation. Alternatively we could estimate what the attribute might be by looking at the frequency of occurrence of a particular data value. Finally, we could use subsequent analysis to predict what the value is likely to be. This form of analysis is referred to a *supervised learning* where a model is constructed based on estimates of similar data points. This approach is often more accurate than just estimating an individual transaction as the supervised approach considers the entire population of transactions.

1.7.2   Data Exploration

Once the data has been cleaned and prepared it is now feasible to start looking for patterns in the data. Even with little experience it is possible to observe linkages between attributes. For instance, in Table 1.1 we can trace the route through manufacture that a particular order takes. We can also see that the `Edge`, `Assembly` and `Paint` workstations have not produced items that need rework. There also appears to be general ranges of cost associated with each of the workstations.

With more data, we can start to pose potential questions such as 'what is the likelihood that the `Cut` operation will produce rework?' The data exploration stage becomes easier with experience and as mentioned previously, is one of the stages where there is a significant benefit of having experience of looking at the data in a particular problem domain.

1.7.3   Model Selection

When people talk about the specific technical skills of a data scientist, they are generally referring to the ability to select algorithms to analyse the data, in order to design a model that can be used for predictions about the future. These algorithms are often referred to as *machine learning* (ML).

There are three broad questions that we want to ask of our data:

1. *'What patterns exist in the data?'*—we need to find potential relationships within the data, beyond those that we observed during the *data exploration* stage. When there is a lot of data, it becomes challenging to identify anything but the most obvious patterns. Experience of a domain helps, but more rigorous enquiry is required. This question is answered by using *unsupervised learning* methods. Unsupervised learning methods identify correlations in data (potential links between attributes such as 'the weather was dry and sunny and ice cream sales increased', or they may group similar attributes together, otherwise referred to as *clustering*.

2. *'What can I predict about the future using the patterns in the data?'*—when we make a forecast, we usually base it on previous knowledge. Forecasting in a business can be quite formal, such as predicting reorder points for stock, or sales volumes for the next quarter. When we have little data to draw upon, humans also make forecasts based upon their experience. We use *supervised learning* methods to make predictions where we have identified a pattern of previous behaviour.

Table 1.2 Common machine learning (ML) algorithms classified by type

Unsupervised learning	Supervised learning	Reinforcement learning
k-means clustering	Decision trees	Multi-armed bandit
Association rules	Naive Bayes classification	
Principal component analysis (PCA)	Support vector machines (SVM)	
Singular value decomposition (SVD)	Random forests	
	Linear regression	
Social network analysis	Ordinary least squares regression	
	Logistic regression	
	Ensemble methods	

3. *'How can I update my predictions based on new knowledge?'*—*reinforcement learning* takes new data as an input to update and refine the output of a prediction. This is different to unsupervised and supervised learning where the models that are generated remain static until they are executed again with a new dataset. Reinforcement learning uses predictions from both unsupervised and supervised approaches to update the current state of an input, which enables 'new' scenarios to be included as part of the model generation, as and when the input data changes.

Table 1.2 gives some examples of common algorithms, categorised by type.

1.7.4 Evaluation

Automation gives us convenience, and it is very useful to have the ability to perform intricate analysis on large sets of data. However, we do need to be careful that we do not implicitly trust that the methods have been applied correctly every time. Our earlier working definition of data science included 'the scientific method', and this means that we should continuously test and verify that our analysis is correct. If we omit this discipline, we may automate disaster.

In the same way that we can compare the relative acceleration performance of a motor car by measuring the time it takes to achieve 60 mph from a standing start, we can use measurements to help us judge the performance of an ML model.

For ML algorithms that classify, we are interested in how accurate the predictions are for a given input dataset. This can be expressed as a percentage of predictions that are correct. For example, the prediction as to whether items required rework from the Cut workstation was 87% correct.

1.8   Do We Need Machine Learning for Industrial Analytics?

With all of the interest in analytics, data science and machine learning, you could be forgiven for thinking that it was all essential. In some ways, to achieve the vision espoused by Industry 4.0, we do need to be implementing cyber-physical systems that can learn from each other. Our current technologies suggest that the use of algorithms to learn about an environment is essential.

But we also need to consider the reality that a good proportion of the industrial environment is faced with. The adoption of ML capabilities requires a certain volume of data that has been prepared and analysed and categorised to some extent already. If we want to use supervised learning approaches to enquire about a process, we need a certain amount of data to train the model, and then we need some more data to test the model.

A common mistake when using ML methods for the first time is to use either the same data to train and test a model, or perhaps to have some overlap in the training and testing datasets. One way to ensure that your model has a high prediction rate is to test it on the same data that it was trained with!

When embarking upon a data science journey for the first time, it can be frustrating to discover that the first bout of analysis appears to produce a set of results that is as difficult to understand as the original data. It is at these times that experience is useful as someone who has practiced data science will be better placed to discern whether the analysis can be altered or if the results are genuinely incomprehensible.

We take the view that we should strive to select the most suitable approach for conducting industrial analytics, and in many cases this might not be the trend of the moment. A lot of positive improvements can be made by the use of simple techniques, and it is wrong for these to be rejected purely on the basis that they are not modern. A common theme in this introductory chapter has been that it is the increased accessibility to computing power that has been the greatest enabler of change in recent years. Analysis tasks that were just too expensive to perform previously are now available either very cheaply or even for free.

Many of these analysis tasks, that can support an organisation's emerging analytics capability, use tried-and-tested, traditional approaches. For the immediate future, this is a reliable and robust path to follow for a lot of industry.

1.9   Learning Activities

? Exercise

1. List the names and provide a brief description for each of the four key stages of analytics maturity for an organisation.
2. What are the key technologies that have enabled analytics activities to be performed inexpensively?

3. Describe the differences between an organisation that uses M2M for data reporting and another that uses IIoT for analytics.
4. How might the industrial Internet challenge traditional ROI assessments?
5. Explain the relationship between IIoT and CPS.
6. Briefly describe the stages of an enquiry that use a data science approach.

? Extension exercise

1. Describe the challenges that an organisation faces when considering the adoption of IIoT. Create an argument for IIoT adoption and describe the stages of implementation you would plan to maximise its impact.
2. Using the working definition of data science, describe what an analytics function might look like for an organisation, including tools, approaches, staff and other resources.
3. Consider a business question that you would like to investigate using data science techniques. Which models would you adopt and what resources would you need? Justify your answer.

References

1. Al-Aqrabi H, Liu L, Hill R, Ding Z, Antonopoulos N (2013) Business intelligence security on the clouds: Challenges, solutions and future directions. In: IEEE seventh international symposium on service-oriented system engineering. IEEE, pp 137–144
2. Al-Aqrabi H, Hill R, Lane P, Aagela H (2020) Securing manufacturing intelligence for the industrial internet of things. Fourth Int Congress Inf Commun Technol 2020:267–282
3. Alsboui T, Qin Y, Hill R, Al-Aqrabi H (2020) Enabling distributed intelligence for the Internet of Things with IOTA and mobile agents. Computing 102(6):1345–1363. Springer Vienna
4. Hill R, Al-Aqrabi H (2020) Edge intelligence and the industrial internet of things. In: Advances in edge computing: massive parallel processing and applications. IOS Press (2020), pp 35, 178

Data, Analysis and Statistics 2

2.1 Introduction

The focus of this chapter is to introduce the essential concepts of data and statistics, along with some thoughts about how we can use statistics to analyse data. If your memories of studying statistics at school were not so good or if you find statistics on Websites, television, social media or newspapers confusing, do not worry as we shall be covering only what is required.

Often, when people find statistics confusing it is not because they cannot understand the subject; it is genuinely the case that the statistics have not been presented properly. The suggestions in this chapter should help you not only use statistics for your own analysis, but you will be able to communicate the results clearly.

2.2 The Need for Analysis and Statistics

When we collect data, it is natural to want to be able to reason about our findings. If the dataset is small—a few data points—we can intuitively draw some conclusions about this data. Generally, humans are quite capable at inferencing potential outcomes from small datasets.

However, as the volume of data points increases we lose the ability to comprehend either the magnitude of the data or the detail that is contained therein. We need to be able to abstract ourselves away from the detail so that we can identify and understand any emerging themes in the data. But, we also need to make sure that our abstract view of the data takes account of the detail, as that detail will shape any potential conclusions that we arrive at.

Statistics enables us to make sense of data by guiding us to summarise outcomes from data points. Used in an appropriate way, statistical methods can assist us to view data in a context, and the context can thus help us to turn raw data into information.

© Springer Nature Switzerland AG 2021
R. Hill and S. Berry, *Guide to Industrial Analytics*, Texts in Computer Science,
https://doi.org/10.1007/978-3-030-79104-9_2

Statistical methods can be applied to data to provide us with an overall understanding of the dataset, and they are a useful way of identifying patterns or trends in data.

It is common for business reports to provide summary statistics for a 'snapshot' of results that is some data at a particular point in time. It can be difficult to assess data if it is presented without a context. The simple addition of some historical data to a report immediately enables a trend in the data to be observed, which can make the understanding of the data much easier.

For example, sales results for some consumer products—like ice cream or gas central heating boilers—vary with respect to the seasonal weather. The general trend for consumers is to consume more ice cream in the summer and to replace gas boilers in the winter when they need them the most.

The use of statistics can also help us qualify a set of data by enabling us to compare different, but related datasets. For instance, we might expect the sale of soft drinks to increase along with ice cream during the warmer weather. If we see a drop in sales for a particular soft drink, it immediately focuses our attention on a potential problem to be solved.

The statistics that we are interested in falls into two categories. *Descriptive statistics* helps us organise and summarise data. Tables and charts are instruments that enable these summaries to be visualised and communicated to different audiences. We summarise the data using a range of approaches.

If the data is numeric, we can look at how the data is arranged. For example, are the values concentrated about the average (*central tendency*)? Are all of the values spread out (*dispersion*)? Is the data concentrated at one end of a particular range (*skewness*)?

The data might not consist of numbers though. In this case we are dealing with *categorical* data (non-numeric), and there are some measures that we can use such as

- *Frequency*—how often does a value appear in the dataset?
- *Percentage*—what fraction of the entire dataset is represented by a value?
- *Proportion*—what is the ratio between different values within the dataset?

The second category is that of *inferential statistics*. We use inferential methods to test for differences between groups of data. This can take the form of hypothesis testing, where we pose a question to be proved or disproved. Inferential statistics allows us to make generalised predictions about an entire *population* by using data from a subset of the population, referred to as a *sample*.

2.3 Qualitative and Quantitative Data

Data that we collect can either be *qualitative* or *quantitative*. Qualitative data represents events or observations that cannot be discretely quantified. For example, the study of psychology requires thoughts, feelings, emotions and opinions to be investigated.

Table 2.1 List of transactions from an organisational database. Rows of data represent *data points*. Columns of data represent *attributes/features/dimensions* (terms used interchangeably)

RecordID	WorkCentre	StartDate	OrderID	Completed	UnitCost	Rework
1	Cut	10-01-2021	10056	10-01-2021	14.35	N
2	Cut	10-01-2021	10073	10-01-2021	22.41	Y
3	Drill	09-01-2021	10021	10-01-2021	8.76	N
4	Edge	09-01-2021	10013	11-01-2021	2.56	N
5	Drill	09-01-2021	10034	11-01-2021	9.65	Y
6	Assembly		10056	11-01-2021	22.88	N
7	Paint		10073	11-01-2021	2.74	N
8	Assembly	10-01-2021	10073	11-01-2021	37.82	N

These do not normally 'map' across into a number, and methods have been developed to assist with this. You may have completed a survey where your 'level of satisfaction' has been converted into a scale with five options, from which you select one, ranging from 'very satisfied' to 'extremely unsatisfied'.

Alternatively, we have quantitative data which represents things or events that can be measured objectively. The number of items produced per hour, an item's weight, the length of an object, the temperature of an individual, etc., are all examples of quantitative data.

Quantitative data can be further broken down into two categories:

- *Discrete* data—these are items that can be counted. The number of products in stock for instance.
- *Continuous* data—measurements that can be recorded to a high level of precision, such as temperature or length. When handling such data, the accuracy (and calibration) of the measuring instrument is important when comparing different datasets.

2.4 Data Terminology

The important terminology of data was introduced in Sect. 1.7.1, but briefly, Table 2.1 summarises the key points. Every row in the table is a collection of *attributes*, and together they constitute a *record*. We usually refer to each row as a *data point* or a record. Each part of the data point might be referred to as a *feature, attribute* or *dimension*.

2.5 Data Quality

Data quality is of prime importance if we are to rely on data for decision-making. Keeping the quality of data in mind means that we must prescribe and maintain a consistent level of effort when collecting, preparing and processing the data.

? Reflection question

Your company has installed remote temperature monitors on each of the refrigerated lorries in a distribution fleet. A few of the monitors appear to have not produced readings for a certain period of time. What are the potential issues for your dataset, and any subsequent decisions that you might take?

The first issue is the fact that there shall be values that are missing from the dataset. The date and times that the data is missing might be a clue that helps identify what an issue is. Data that is missing often tells us more than we think.

Missing data leads to inconsistency. If we can identify inconsistency in the data, what might that tell us about the performance of a system?

Is this a faulty sensor or upload link?

Was the monitoring unit turned off for some reason?

Was there a power failure?

When we use data for analysis, we are usually looking for patterns and trends. While inconsistencies can seem like a problem, they can often help us understand a system better.

Duplicated data can be generated by systems. Sometimes, a monitoring unit can rebroadcast/transmit data if there has been a problem with a network link. It might be that the unit has been rebooted and it has resent the last chunk of data that was recorded. The system time of the monitoring unit might have been reset.

Accuracy is something that may be of concern. How can we assure ourselves that 5 °C in one refrigerated container is the same as 5 °C reported by another lorry?

How can we be sure that the sensors report the readings accurately across a range of values?

System time is one piece of data that is notoriously challenging to manage across separate embedded systems. If we have an incident that is time critical, we need to be able to observe what happened when across a range of devices within the system.

All of these factors lead us to conclude that discrepancies in data have a disproportionately greater effect as the size of the dataset reduces which leads us to look for more data; as mentioned in earlier chapters, the emergence of cloud computing facilitated the provision of computing infrastructure that enables us to process and store ever-increasing volumes of data. Many of the statistical techniques that we utilise require a minimum volume of data for them to work correctly.

As we become more data-literate, we start to understand what a data-rich environment might look like. Familiarity with the use of tools and techniques allows us to gain the experience that we need to conduct robust analysis. There are techniques that are constantly being developed to deal with *small data* problems, though there is till a limit to what can be inferred from less, rather than more, data.

For example, it would be unwise to predict the weather purely from a dataset of temperatures recorded at lunchtime each day; we therefore need to consider what data we have available and then what data we might need to gather, to make a believable forecast of a future state.

Figure 2.1 shows a dataset that contains a number of potential data quality issues. Try and discover what they are, using the preceding text to help. Once you have created a list of problems, look at Fig. 2.2 to see if you found all of them.

When you have identified the issues, try and think about the potential impact that these data issues might have on the outcome of some analysis. What could you do to minimise or mitigate such problems?

2.6 Scales of Measurement

When we talk about data and especially when we need to communicate our analytics to others, it is important that we are able to define each of the variables in a way that can be understood. Numbers are usually understood as to be part of a *scale*. A scale allows us to define the relationship between two values [2].

Depending on the nature of that relationship, we can use that understanding to interpret the data in different ways. Having an understanding of the scale enables us to apply the correct approach to interpreting the data.

There are four scales of measurement as follows:

1. Nominal
2. Ordinal
3. Interval
4. Ratio.

2.6.1 Nominal Data

If data is discrete and 'named', we consider the data to be *nominal*. An example of nominal data is eye colour: brown, blue, grey, green, etc. The names enable us to *categorise* the values, but there is no indication of any *quantity*. A person's eye colour cannot indicate size for instance.

In some cases there may be only two categories of nominal data. An individual may have an infection, or they may not. From a data perspective we say that these values are *binary* in that they are either of value {1} (has an infection) or {0} (does not have an infection).

Index	Time	Name	DOB	Age	Gender	Ticket	Fare
0	05:54:12	Allen, Miss. Elisabeth Walton	19/02/1952	67	female	24160	211.3375
1	07:13:06	Allison, Master. Hudson Trevor	26/11/1959	59	male	113781	151.55
2	03:30:41	Allison, Miss. Helen Loraine	06/18/1950	69	female	113781	151.55
3	08:34:12	Allison, Mr. Hudson Joshua Creighton	21/07/1974	45	male	113781	151.55
4	04:04:20	Allison, Mrs. Hudson J C	10/04/1954	65	female	113781	151.55
5	08:54:19	Anderson, Mr. Harry	13/11/1961	57	male	19952	-26.55
6	04:50:22	Miss. Andrews, Kornelia Theodosia	07/02/1959	60	female	13502	77.9583
7	07:14:20	Andrews, Mr. Thomas Jr	12/12/1982	36	male	112050	0
8	02:16:17	Appleton, Mrs. Edward Dale	16/12/1972	46	female	11769	51.4792
9	08:17:54	Artagaveytia, Mr. Ramon	23/06/1974	45	male	PC 17609	49.5042
10	03:40:01	Astor, Col. John Jacob	13/03/1996	23		PC 17757	227.525
11	10:32:11	Astor, Mrs. John Jacob	13/10/1958	61	female	PC 17757	227.525
12	18:02:56	Aubart, Mme. Leontine Pauline	08/01/1953	66	female	PC 17477	69.3
13	07:43:32	Barber, Miss. Ellen "Nellie"	29/03/1953	66	female	19877	78.85
14	09:24:11	Mr. Barkworth, Algernon Henry Wilson	15/03/1960	59	male	27042	30
15	04:35:36	Baumann, Mr. John D	17/04/1952	67	male	PC 17318	25.925
16	09:57:39	Baxter, Mr. Quigg Edmond	17/05/1975	44	male		247.5208
17	05:56:47	Baxter, Mrs. James	29/03/1990	209	female	PC 17558	247.5208
18	02:16:17	Appleton, Mrs. Edward Dale	16/12/1972	46	female	11769	51.4792
19	09:55:45	Bazzani, Miss. Albina	05/06/1979	40	female	11813	76.2917

Fig. 2.1 Sample dataset. What potential challenges can you discover?

Index	Time	Name	DOB	Age	Gender	Ticket	Fare
0	05:54:12	Allen, Miss. Elisabeth Walton	19/02/1952	67	female	24160	211.3375
1	07:13:06	Allison, Master. Hudson Trevor	26/11/1959	59	male	113781	151.55
2	03:30:41	Allison, Miss. Helen Loraine	06/18/1950	69	female	113781	151.55
3	08:34:12	Allison, Mr. Hudson Joshua Creighton	21/07/1974	45	male	113781	151.55
4	04:04:20	Allison, Mrs. Hudson J C	10/04/1954	65	female	113781	151.55
5	08:54:19	Anderson, Mr. Harry	13/11/1961	57	male	19952	-26.55
6	04:50:22	Miss. Andrews, Kornelia Theodosia	07/02/1959	60	female	13502	77.9583
7	07:14:20	Andrews, Mr. Thomas Jr	12/12/1982	36	male	112050	0
8	02:16:17	Appleton, Mrs. Edward Dale	16/12/1972	46	female	11769	51.4792
9	08:17:54	Artagaveytia, Mr. Ramon	23/06/1974	45	male	PC 17609	49.5042
10	03:40:01	Astor, Col. John Jacob	13/03/1996	23		PC 17757	227.525
11	10:32:11	Astor, Mrs. John Jacob	13/10/1958	61	female	PC 17757	227.525
12	18:02:56	Aubart, Mme. Leontine Pauline	08/01/1953	66	female	PC 17477	69.3
13	07:43:32	Barber, Miss. Ellen "Nellie"	29/03/1953	66	female	19877	78.85
14	09:24:11	Mr. Barkworth, Algernon Henry Wilson	15/03/1960	59	male	27042	30
15	04:35:36	Baumann, Mr. John D	17/04/1952	67	male	PC 17318	25.925
16	09:57:39	Baxter, Mr. Quigg Edmond	17/05/1975	44	male		247.5208
17	05:56:47	Baxter, Mrs. James	29/03/1990	209	female	PC 17558	247.5208
18	02:16:17	Appleton, Mrs. Edward Dale	16/12/1972	46	female	11769	51.4792
19	09:55:45	Bazzani, Miss. Albina	05/06/1979	40	female	11813	76.2917

Fig. 2.2 Data issues illustrated with highlighted areas

2.6.2   Ordinal Data

Sometimes the data appears to be categorical, but there is also a clear or ranking that means we can arrange the data in order.

An Olympic athlete can receive a bronze, silver or gold medal if their performance in competition is either third, second or first place. While we can rank the data, we cannot say how far apart the values are. For the case of our Olympic event, if we assume that it is a race, we are only concerned with the individual race times to put them in order from fastest to slowest.

If there is a 2 s cap between the first and second place (gold and silver medal), it does not matter what the time is between the second and third places is, as long as the bronze medal is awarded to the *next fastest* time.

This means that the time difference between the race finishing times does not have to be the same (it would be unlikely as well), and therefore we cannot infer that the distance between values is the same.

? Reflection question

At some time you have probably been asked to complete an opinion survey. Often, each question is followed by a series of numbers, say 1 to 5, where you indicate your opinion on a scale of values from 'very satisfied' to 'very dissatisfied'. What type of data is being recorded?

What issues do you see with collecting data in this way?

2.6.3   Interval Data

If we can quantify the difference between two values on a scale then we say that the data is *interval data*. On a Farenheit temperature scale, the interval from 65–70° is the same difference as from 105–110°. The scale is *continuous*, and each numerical item of data is a standardise distance from the next value.

However, the placement of 'zero' is arbitrary. Some interval data scales may include negative values for instance—temperature—but not length. The time on a clock is another example of interval data. We can calculate the differences between two values, but we cannot multiply or divide interval data scales.

2.6.4   Ratio Data

Ratio data is similar to interval data except that we must have an absolute zero value. If we measure weight in kilograms a value of zero means no weight. We also know that 50 kg is twice as heavy than 25 kg, which means that we can multiply and divide values within ratio data scales. Thus, scales of measurement can be represented as follows:

1. *Nominal*—data that contains a named variable;
2. *Ordinal*—data that contains a named variable and is ordered;
3. *Interval*—data that contains a named variable and is ordered in proportionate intervals;
4. *Ratio*—data that contains a named variable and is ordered in proportionate intervals, with an absolute value.

? Reflection question

Test your understanding with the following. Which scales are the best for the different data attributes?

- Time—seconds/hours/days/years;
- Weekdays—Monday, Tuesday, Wednesday, etc.;
- Temperature—Celsius;
- Length—mm/inches;
- Temperature—Kelvin;
- Hair colour—blonde, red, brunette, etc.

2.7 Central Tendency

A useful understanding of a collection of data is to understand where data might be clustered. We use this understanding when we calculate an average—usually the *mean*—of a set of values. Other measures of central tendency are *median*, which is the middle value after all the values have been arranged in order from smallest to largest, and the *mode*, which represents the most frequently occurring value in a dataset.

The measure that we use depends on the type of variable, plus the *shape* of the distribution of the dataset.

2.7.1 Mean

The mean is calculated by adding up all of the values in a dataset and then dividing by the number of values in the dataset. For example, if a business process is monitored over 5 days and the times recorded for each day give the following dataset: {4.5, 4.9, 5.3, 4.4, 4.9}, then the mean is

$$\text{mean} = \frac{(4.5 + 4.9 + 5.3 + 4.4 + 4.9)}{5} = 4.8$$

? Reflection question

Consider the following two datasets:

1. $\{1, 97, 3, 99\}$
2. $\{48, 53, 52, 47\}$

What is the mean for each of the datasets?

In the reflection question, the *spread* of the data is very different, even though the means are identical. In fact, the first dataset is spread out more than the second. Understanding this can help us interpret data. We may have *outlier* data; these are spurious values that are a distance from the bulk of the values. They affect the mean and thus can affect any conclusions that we draw. Therefore, we need a statistical approach to help us visualise *dispersion*. The mean is not suitable for nominal or ordinal data, only interval/ratio data.

2.7.2 Median

The median is the central value of an ordered set of values. For our set of process times $\{4.5, 4.9, 5.3, 4.4, 4.9\}$, when rearranged in order they become $\{4.4, 4.5, 4.9, 4.9, 5.3\}$. The central value is $\{4.9\}$. If the set of values is an even number, we take the mean of the central two values.

Unlike the mean, a median is affected less by outlier values. It is sometimes referred to as the *50th percentile*. The median can be used for ordinal data in some cases, but it is mostly used with interval/ratio data.

2.7.3 Mode

For our dataset of process times $\{4.5, 4.9, 5.3, 4.4, 4.9\}$, the most frequently occurring value is $\{4.9\}$ so this is the mode. The mode is robust and is not affected by extreme values in a dataset. Not every dataset has a mode, and conversely there can be several modes in a dataset. The mode is most suitable for categorical or discrete data, but it can be used for nominal, ordinal and interval/ratio data also.

? Reflection question

Using the sample data in Fig. 2.3, which measure of central tendency would be the correct one to use?

Age	Salary (K)	Hair Colour	Highest Qualification	Date
24	28	Brown	BSc	10/03/2019
45	32	Red	BSc	15/04/2019
28	20	Blonde	MSc	22/09/2019
37	25	Blonde	PhD	13/06/2019
52	26	Black	MSc	17/12/2018
63	45	Grey	BSc	29/11/2018
43	30	Brown	BSc	03/08/2019
46	54	Brunette	PhD	15/06/2019
58	110	Grey	MSc	11/08/2018
31	23	Blonde	BSc	31/10/2019

Fig. 2.3 Measuring the central tendency of a sample dataset

a) Values spread out giving **high** dispersion

b) Values clustered together giving **low** dispersion

Fig. 2.4 Visualising the spread of data with the same mean

2.8 Dispersion

As discussed earlier, measures of central tendency do not describe the spread of data, or how it is *dispersed*. Figure 2.4 illustrates two datasets with identical means, where the data points are spread differently. The greater spread of data in (a) is dispersed higher than that in (b). Since dispersion is an important factor when looking at data, we tend to report dispersion together with measures of central tendency.

It is common for business functions to report only measures of central tendency—usually the mean—without illustrating the dispersion of the data. This can be misleading and is not good practice. Several measures of dispersion that are useful are as follows:

- range;
- interquartile range;
- variance;
- standard deviation.

Fig. 2.5 Interquartile range (IQR)

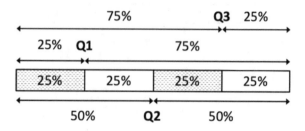

2.8.1 Range

Range is the difference between the largest and smallest value in a dataset. For our process data, {4.5, 4.9, 5.3, 4.4, 4.9}, the range is {0.5}. Since the range only takes into account the extremities of a dataset, it is influenced by outlier values.

2.8.2 Interquartile Range

The interquartile range (IQR) separates an ordered data set into four parts. Figure 2.5 illustrates the various proportions. 25% of the values are below the first quartile (Q1), and 75% are above it. The second quartile (Q2) has 50% of the values below it, and 50% above it, so this is the same as the *median*. Quartile 3 has 75% of the values below it, and 25% above it.

If we consider the following dataset: {42, 43, 44, 44, 50, 52, 56, 57, 61, 61}, the median (and Q2) is

$$\text{Median} = \frac{50 + 52}{2} = 51$$

Q1 = 44 and Q3 = 57. The interquartile range (IQR) is

$$\text{IQR} = \text{Q3} - \text{Q1} = 57 - 44 = 13$$

2.8.3 Variance

So far we have seen simple measures of dispersion that are useful for quick assessments of data. They are easy to calculate, but they are still susceptible to undue influence from outlier data. The IQR resists this to some extent, but it is still evident.

Variance is a measure of the distance of each value from the mean. The measure is expressed as the average of the *squared* deviations from the *mean*.

We calculate it as follows:

$$\text{Variance, } \sigma^2 = \frac{\Sigma(x_i - \bar{x})^2}{N} \tag{2.1}$$

Table 2.2 shows how we calculate the values.

Table 2.2 Calculating the variance for a dataset

Data point, x	Distance between the data point x_i and the mean, $\bar{x}$ $x_i - \bar{x}$	Distance squared $(x_i - \bar{x})^2$
5	−6	36
3	−8	64
12	1	1
8	−3	9
15	4	16
18	7	49
5	−6	36
22	11	121
	Sum	332

$$\text{Variance, } \sigma^2 = \frac{332}{8} = 41.5$$

Thus, the variance illustrates the dispersion of a dataset. The larger the variance, the more spread out the data is.

2.8.4 Standard Deviation

While the variance gives us a measure of dispersion, the value is squared which might not make sense depending on the units that we are using. For instance, if we used our process times, the variance would calculate the square of the time which is meaningless. The standard deviation addresses this by taking the *square root* of the variance to show the deviation from the mean, in the same units as the dataset being assessed.

$$\sigma = \sqrt{\frac{\Sigma (x_i - \bar{x})^2}{N}} \tag{2.2}$$

Using the data from Table 2.2, we simply take the square root of 41.5:

$$\text{Standard deviation, } \sigma = \sqrt{\frac{332}{8}} = 6.4$$

2.8.5 Frequency

We can look at the number of times a particular value occurs in a dataset and then use that to present the data. Table 2.3 illustrates how the left hand set of values can be transformed by counting the occurrence of each of the unique values. If the dataset contains lots of individual values then the frequency plot may not transform the data much as the count for each value tends towards 1.

This can be addressed by considering the values as intervals, commonly referred to as 'bins'. Table 2.4 shows how the counts change by representing the same data as intervals prior to counting. As a check, the sum of the counts must equal the number of data points in the dataset.

2.8.5.1   Relative Frequency

Now that we have a frequency count we can represent the *relative frequency* as a proportion of the overall count.

$$\text{Relative frequency} = \frac{\text{Class frequency}}{\text{dataset size}} = \frac{f}{N}$$

Table 2.5 demonstrates the relative frequencies for each of the class bins.

2.8.5.2   Cumulative Frequency

Table 2.6 demonstrates how we calculate the *cumulative frequency* by adding the frequency count for each bin to the previous counts.

Table 2.3 Representing data by counting the number of occurrences in a dataset

19	29	34	39	46	50	Class	Frequency
20	31	36	40	47	53	19	1
21	31	37	40	47	55	20	1
23	32	38	41	48	55	21	1
28	32	38	42	48	56	23	1
						28	1
						29	1
						31	2
						32	2
						34	1
						36	1
						37	1
						38	2
						39	1
						40	2
						41	1
						42	1
						46	1
						47	2
						48	2
						50	1
						53	1
						55	2
						56	1

Table 2.4 Arranging the data into intervals (bins) before counting the frequency of occurrence

						Class interval	Frequency
19	29	34	39	46	50		
20	31	36	40	47	53	19 to 26	4
21	31	37	40	47	55	27 to 34	7
23	32	38	41	48	55	35 to 42	9
28	32	38	42	48	56	43 to 50	6
						51 to 58	4
						Sum	30

Table 2.5 Calculating the proportion of a dataset (N) that is of a relative frequency

						Class	Frequency	Relative frequency
19	29	34	39	46	50			
20	31	36	40	47	53	19 to 26	4	0.133
21	31	37	40	47	55	27 to 34	7	0.233
23	32	38	41	48	55	35 to 42	9	0.300
28	32	38	42	48	56	43 to 50	6	0.200
						51 to 58	4	0.133
						Sum	30	1

Table 2.6 Calculating the cumulative frequency by successively adding a class frequency to all of the previous class frequency counts

						Class	Frequency	Cumulative frequency
19	29	34	39	46	50			
20	31	36	40	47	53	19 to 26	4	4
21	31	37	40	47	55	27 to 34	7	11
23	32	38	41	48	55	35 to 42	9	20
28	32	38	42	48	56	43 to 50	6	26
						51 to 58	4	30
						Sum	30	

2.9 Histogram

Numeric data can be visualised as a *histogram* as per Fig. 2.6. The interval data (bins) are represented along the x-axis, and the frequency count is plotted on the y-axis.

Histograms are often confused with *bar charts*, but this is incorrect. Histograms represent the distribution of variables, whereas bar charts show *categorical data*. As such, the bars of a bar chart may be reordered, but the intervals of a histogram cannot.

Since the width of a bar within a histogram corresponds to an interval, the widths can vary or even be zero if no data exists in that interval. The volume of data within an interval is represented by the *area* of a bar.

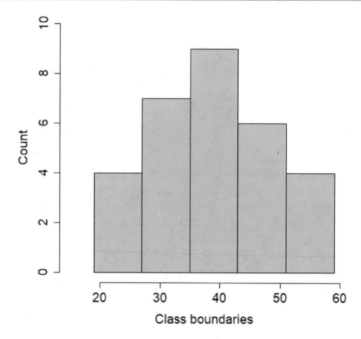

Fig. 2.6 Visualising interval data with a histogram

2.9.1 Cumulative Frequency Graph

Figure 2.7 demonstrates the line plot when the cumulative frequency is the x-axis and the upper boundary of the interval is the y-axis.

Figure 2.8 shows the median (50th percentile, Q2) superimposed onto the cumulative frequency plot. The remaining quartiles are shown in Fig. 2.9. To calculate the IQR, we subtract the lower quartile (Q1) from the upper quartile (Q3).

2.10 Shape of the Data

Plotting the statistical functions as applied to data results in different plots. These plots demonstrate a *shape* that we can use to make an assessment of the data. With practice and experience you shall be able to 'see' the shape in your mind by interpreting the statistical results, without having to plot the data. This is an example of how we can develop statistical 'intuition' and makes the evaluation of data much quicker. The shapes encountered are typically categorised as follows:

- symmetrical;
- uniform;

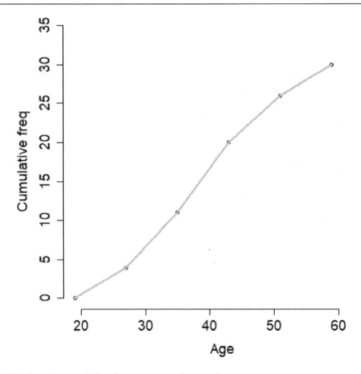

Fig. 2.7 Plotting the cumulative frequency as a line graph

- asymmetrical (skewed);
- bimodal (or multi-modal if more than two instances).

We can also affect the shape of the plot by altering the resolution of the intervals. As the bin size increases (the interval becomes larger) the bar widths increase, and the overall profile of the plot becomes more 'jagged'. Conversely, a reduction in bin size leads to more discrete bins, and a smoother shape emerges (Fig. 2.10). A *probability density function* is a way of indicating the likelihood that a particular value might occur. The shape that is plotted is the same as that for a frequency distribution; instead of plotting the frequency count on the y-axis, we plot the relative *density* on the vertical axis. Figure 2.11 shows the probability density function curve.

2.10.1   Normal Distribution

The normal distribution (sometimes referred to as a 'bell curve' on account of its shape) is a probability density function where the values are symmetric around the mean (shown in Fig. 2.12). This means that the occurrences nearer the mean are more likely to happen than those further from the mean (in the 'tails' of the curve).

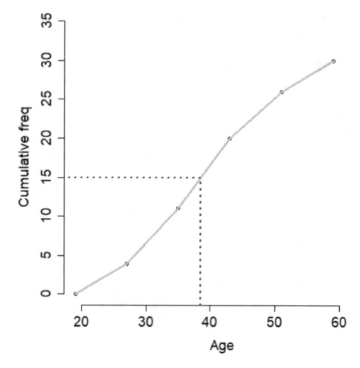

Fig. 2.8 Cumulative frequency line plot, with the median value illustrated

Many natural occurrences, when plotted as a probability density function, resemble the normal distribution. Blood pressure and intelligence are two examples. The essential characteristics of a normal distribution are as follows:

- The mean value is equivalent to the median *and* the mode.
- The maximum height of the curve is always at point μ (the mean).
- 68.3% of the data is contained within the area of one standard deviation either side of the mean = $\mu \pm 1\sigma$.
- 95.4% of the data is contained within the area of two standard deviations either side of the mean = $\mu \pm 2\sigma$.
- 99.7% of the data is contained within the area of three standard deviations either side of the mean = $\mu \pm 3\sigma$.

If you have experience of Six Sigma for leading process improvement, you will recognise the last item. Six Sigma represents six standard deviations, which is three either side of the mean, giving 99.7% of the values in a dataset. Usually, any tolerance for defects is only considered acceptable if it is outside of the six standard deviations.

If we are confident that our data distribution will fit this curve, we can make predictions about the future based on this curve. However, not all datasets can be

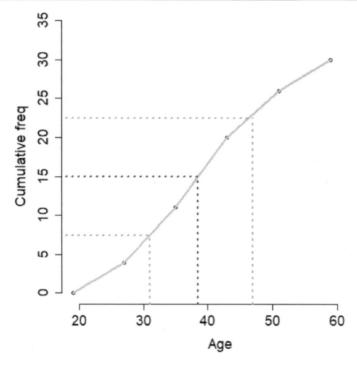

Fig. 2.9 Cumulative frequency line plot, with upper and lower quartiles illustrated

represented by a normal distribution so it is necessary to understand the conditions necessary for a normal distribution to be useful.

? Reflection question

From the preceding explanations, how would you ascertain whether a dataset was distributed normally?

The first thing to try is to plot the data. Does it look like a normal distribution shape? The curve should be symmetrical about the mean and has only one peak (the mean).

We can now calculate the descriptive summary statistics to find the mean and standard deviation. Does approximately 95% of the data lie within ±2 standard deviations of the mean? If so, does 68% lie within ±1 standard deviation of the mean?

As the variance within a distribution increases we know that the data is more spread out. This has the effect of flattening the normal distribution curve illustrated in Fig. 2.13.

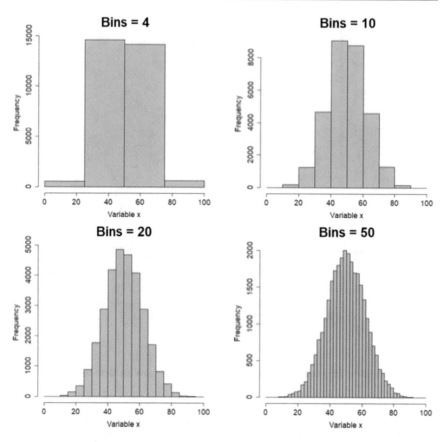

Fig. 2.10 Illustration of how bin size affects the shape of the plot of frequency

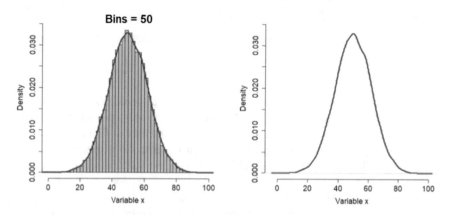

Fig. 2.11 Illustration of how bin size affects the shape of the plot of frequency

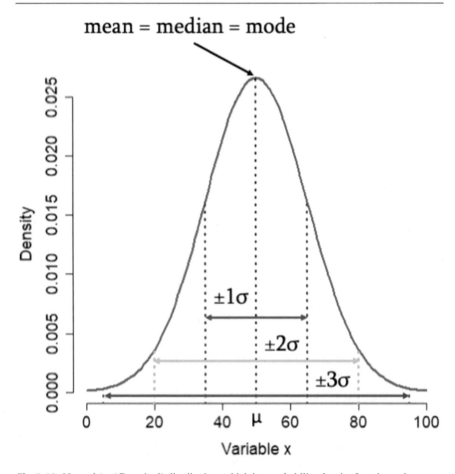

Fig. 2.12 Normal (or 'Gaussian') distribution, which is a probability density function, whose area under the curve = 1

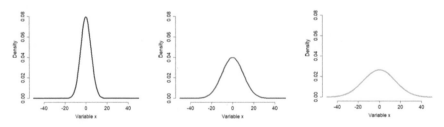

Fig. 2.13 As the variance increases, the shape of the normal distribution becomes flatter

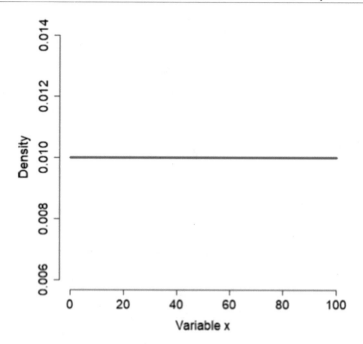

Fig. 2.14 Uniform distribution has no mode

2.10.2 Uniform Distribution

A uniform distribution has a mean, a median but not a mode as all the values have the same frequency (Fig. 2.14). In practice, this is a useful distribution to know about if we do not know which distribution is most suitable. Often referred to as 'the distribution of general ignorance', if we need to model a process where there is no pre-existing data, we start with a uniform distribution.

2.10.3 Bimodal Distribution

So far the distributions have had a maximum value that either represents a peak for a normal distribution or the central value of a uniform distribution. However, distributions can have multiple *modes* and are thus bi (two) or multi-modal, as per Fig. 2.15. Multi-modal distributions can infer that there is clustering of the data, and it might be that there are distinct phenomena that are being observed. For instance, examination performance can often be represented by a bimodal distribution; students can often cluster around a high mark, as well as a pass mark. The emergence of a multi-modal distribution generally stimulates further data collection and investigation.

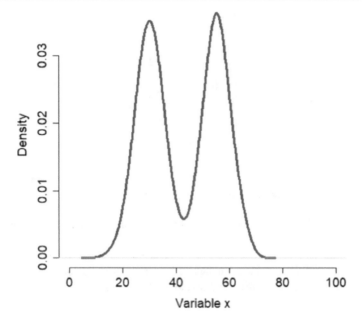

Fig. 2.15 Probability density function with two modes (bimodal)

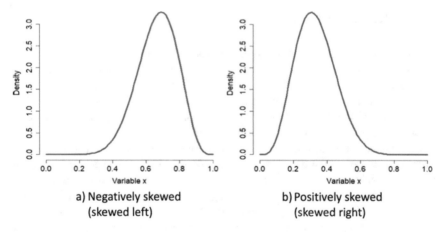

Fig. 2.16 Probability density function with two modes (bimodal)

2.10.4   Skewed Distributions

Distributions that are *assymetrical* have a longer tail to the left or the right of the mean. Figure 2.16 illustrates the shapes. Each distribution is described by the side where the long tail is present. A long tail on the right side means that the distribution is described as *positively skewed* or *skewed right*. Alternatively a tail to the left means that the distribution is described as *negatively skewed* or *skewed left*.

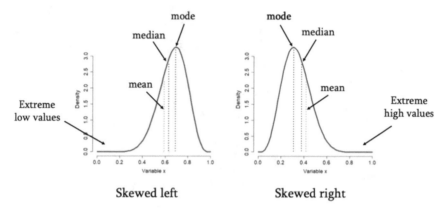

Fig. 2.17 Skewed distributions showing the relative position of the mean, median and mode

For a given distribution, the mean lies on the same side as the long tail, on the left of a negatively skewed distribution and on the right of a positively skewed distribution (Fig. 2.17).

2.11 Visualising Data

The use of statistical functions helps us describe and understand data better. We have also seen that while we can use these functions to calculate descriptive statistics, when these functions are plotted as graphs it is easier to understand the 'shape' of the data. In data-rich environments, the accurate comprehension and communication of data is important. In this modern era of data-producing functions, the correct use of data to inform decision-making is even more crucial.

Data visualisation has become an area of considerable interest as organisations struggle to interpret their data. The topic straddles the technical aspects of preparing data, the psychological comprehension and perception of data, and how we place data in context and interpret it. As such, data visualisation techniques are advancing rapidly and becoming ever more sophisticated.

Remaining true to the spirit of this book, we take the approach that produces the maximum results for the minimum of effort. This means that the approaches are straightforward to apply, and more importantly, they will get used. It does not take long to find poorly presented data, whether it be within an organisation or in the public media such as television and the World Wide Web. The following guide will help the unwary to avoid many of these transgressions.

Perhaps one of the principal aspects to remember is that different data visualisation approaches have different limitations. Spreadsheet tools such as MS Excel provide a wide range of visualisation types, and it can be tempting to consider (even subconsciously) the one that is most aesthetically pleasing. However, good use of the basics

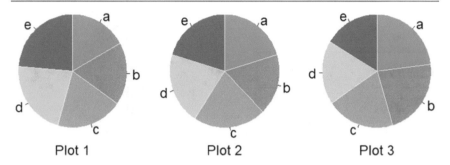

Plot 1   Plot 2   Plot 3

Fig. 2.18 Three pie charts

will generally ensure that you can (a) understand your data and (b) communicate your findings in a way that maximises its comprehension.

Data visualisation is often used in two ways. First, we use visualisation so that we can try and make sense of our analysis as we are doing it. We might use a visualisation as we prepare sensor data, for example, to see where the outliers are and to decide as to whether we include them or not.

Second, we use data visualisation to communicate our work to others. This might stimulate further questions that in turn will generate more analysis to be performed.

This distinction is important as the audience is generally different. When we use visualisation as part of the analysis, we do not have to consider the needs of the audience in the same way that we should when we are presenting the results to others.

Simplicity should be an objective when preparing a visualisation. What does the reader need to conclude from the visualisation?

Reducing a comprehensive visualisation into a series of separate visualisations that effectively communicate a restricted number of points can help lessen the cognitive load and that usually improve the chances of your point being made successfully.

We shall now consider each of the visualisations in turn:

- pie charts;
- bar charts;
- line charts;
- scatter plots.

2.11.1   Pie Charts

The favourite of many a business presentation (especially the three-dimensional version), pie charts are used to compare proportions or percentages. Each proportion represented as a slice of the 'pie'. Figure 2.18 shows three pie charts. The key question here is that while there appears to be some variation in the sizes of the segments, what meaningful comparison can be made?

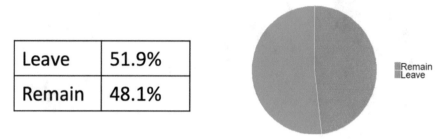

Leave	51.9%
Remain	48.1%

Fig. 2.19 Two slices of similar proportion

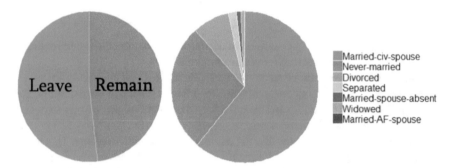

Fig. 2.20 As the number of slices increases, a legend aids comprehension

We know that the whole pie represents 100%, but there are too many similar segments to work out what the proportions might be. The psychologists tell us that humans are generally not that accurate at estimating angles, so most people tend to divide the whole into a number of parts and then approximate from there. Once there are more than four segments, many people find this challenging [1].

Conversely, fewer slices (Fig. 2.19) also suggest that the pie chart is not really communicating anything that could not be deduced from the raw data. Fewer slices can usually be labelled on the pie chart itself, but as the number of slices grows it is good practice to use a legend as in Fig. 2.20.

When using multiple pie charts to demonstrate a comparison, it is preferable to have the same 'starting point' for each chart, as a point of reference. We do this by ordering the slices from largest to smallest, with the first (largest) slice commencing at 12 o'clock, as in Fig. 2.21.

If a pie chart contains a lot of slices, but there are some which are more substantial, with several that are relatively small (Fig. 2.22), then it can help to group the smaller slices into a larger category such as 'other'. To summarise, we should remember that humans find angles difficult to interpret, and we should restrict the number of categories (that become slices) within the chart. As a result, the use of pie charts becomes quite restricted, and some data visualisation professionals advocate other visualisation approaches instead, one of which is the bar chart.

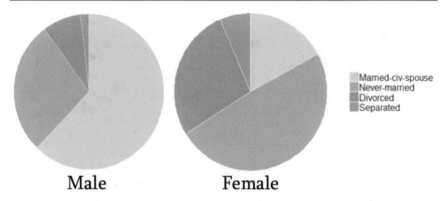

Fig. 2.21 Orient the pie chart so that the largest slice starts at 12 o'clock and proceeds clockwise

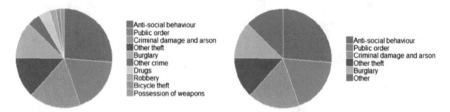

Fig. 2.22 Lots of small slices become unreadable, so it is perhaps better to group the smaller slices into a broader category

2.11.2 Bar Charts

The bar chart is widely understood and illustrates a relationship between categories (the x-axis) and quantitative values (the y-axis). Variables can be ordered (ranked), and a bar chart is a sensible alternative to the pie chart.

Figure 2.23 illustrates an advantage of bar charts over pie charts; the differences between the categories are much more evident, as is the ranked order of them. This understanding is therefore obscured by the use of a pie chart as a visualisation, even though the insight lies within the same data.

? Reflection activity

Look at the bar chart in Fig. 2.24. List the ways in which it could be enhanced to aid comprehension.

Your list of enhancements should contain some or all of the following:

- X-axis text labels are illegible.
- Y-axis text labels are illegible.

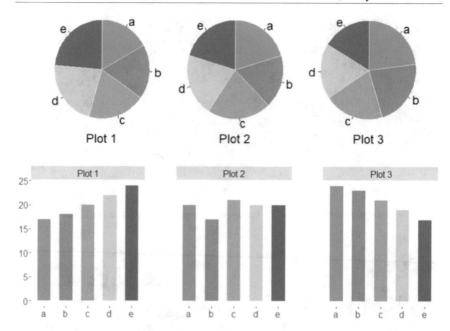

Fig. 2.23 Lots of small slices become unreadable, so it is perhaps better to group the smaller slices into a broader category

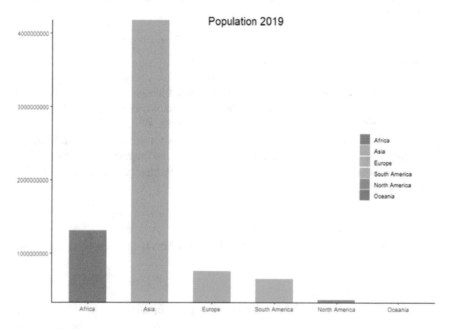

Fig. 2.24 Bar chart with problems

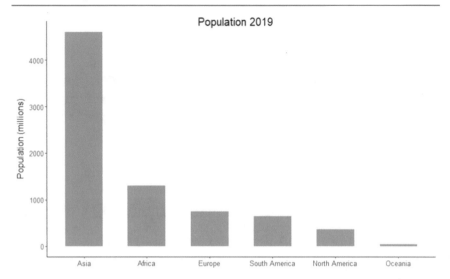

Fig. 2.25 Bar chart after enhancement

- No units on the axes.
- No indication of zero on the y-axis.

Some other questions that we could pose are as follows:

- Is the legend required? Could it be removed?
- Is there missing data for the last category?
- Do the bars need to be different colours?

   If we address the issues above, we can see the changes in Fig. 2.25.

2.11.3   Line Charts

Businesses are often interested in monitoring changes over time, especially when they are investigating performance. A line chart is commonly used for illustrating the *trend* of some data over time (Fig. 2.26). Since we are investigating a trend, it is deemed acceptable to truncate the y-axis (start above zero) so that the trend is maximised visually.

   Multiple lines can be superimposed to indicate relative performance, but good practice is to limit the number of lines to a maximum of five.

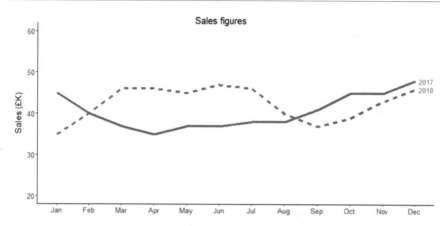

Fig. 2.26 Line chart with two trends. Note that the y-axis starts just below 20

2.11.4   Scatter Plots

Amongst the charts discussed so far, scatter plots are generally received as appearing the most 'technical', and some audiences are intimidated by them. This needs to be considered when thinking about the most appropriate visualisation to communicate your results.

Scatter plots take two quantitative variables and plot them against each other. As the points are plotted, a trend may emerge. Scatter plots are a common method of monitoring process performance in factories. They are typically called 'control charts', and they are used to record parameters while a machine is running. When a parameter starts to fall outside of some predetermined limits, the operator intervenes to correct the process.

Correlation is a measure of the strength of a linear relationship between the two variables that are being plotted. If an increase in one variable causes an increase in the other variable (and vice versa), there is said to be a *positive* correlation. If the increase in one variable causes a decrease in the other variable (and vice versa), then there is a *negative* correlation. Figure 2.27 illustrates this.

Care needs to be taken when observing correlations. The correlation is an implied relationship between two phenomena, and there could be other reasons (that are not being monitored) that create the appearance of a relationship.

2.12   Learning Activities

? Exercise

1. Describe the two types of quantitative data.
2. List as many data quality issues as you can think of. For each one, explain what the potential impacts of these issues might be.

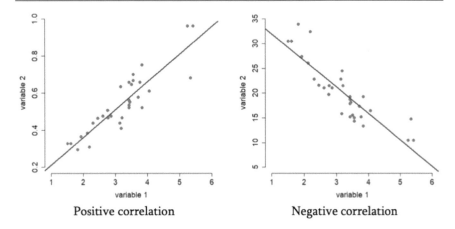

Positive correlation   Negative correlation

Fig. 2.27 Two scatter plots showing positive and negative correlation

3. Explain what the different scales of measurement are.
4. Describe how you would calculate the interquartile range of some data.
5. What is the difference between variance and the standard deviation of a dataset?
6. What does the standard deviation tell you about some data?
7. List the reasons for not using a pie chart as a visualisation technique.

? Extension exercise

1. Think about the descriptive statistics that you can use to assess a dataset. Write down a process to follow that would enable you to identify some important characteristics that will be useful for subsequent analysis.
2. Find a dataset to analyse. This could be from a process at your place of work, from a database or from a freely available dataset. Use the techniques and methods to analyse the data. What have you learned from the data and the analysis as a result?

References

1. Kirk A (2019) Data visualisation, 2nd edn. SAGE Publications
2. Illowsky B, Dean SL, Illowsky B (2016) Introductory statistics. OpenStax College and Rice University

Measuring Operations

3

3.1 Introduction

Managing the operations of a business is both complex and interesting. Experienced managers can often 'look at the numbers' when they prepare to take an operational decision. This skill takes time to develop, especially when it is learned 'on the job'.

There is a faster way to acquire this knowledge, and managers, team leaders or supervisors usually find that if they take an analytics-based approach, they also make better decisions. Sometimes the facts lead to a counter-intuitive solution that is more effective than a decision based purely on experience.

Keeping to the *just enough* mathematics approach of this book, we shall explore ways in which we can use data to calculate measurements that we can factor into our decisions about operations.

The subject of operational research is a mature topic for academic study, and some of it is very detailed and sometimes, rather esoteric. This detail is useful when diagnosing very complex issues in operations management; it does assume though that the more basic analysis has already taken place and any resulting actions have been implemented successfully.

Since we want to help establish and implement the essential aspects of industrial analytics, we shall avoid the detail and concentrate on proven, practical methods that work. The results might not be optimal, but they will demonstrate tangible improvements when applied to a real-world scenario.

Once you develop a proficiency in these approaches you may decide to explore more specific techniques. It is likely though that you shall have achieved your objective to improve operational efficiency and if this is the case, there is no need for further study.

The equipment required for this chapter can be as simple as a pencil, eraser and paper. However, the spreadsheet is now universally accessible and also offers the advantage of being able to be used to dynamically model a situation—asking 'what if?' questions—which is a fundamental part of how we increase our understanding of the key concepts.

© Springer Nature Switzerland AG 2021
R. Hill and S. Berry, *Guide to Industrial Analytics*, Texts in Computer Science,
https://doi.org/10.1007/978-3-030-79104-9_3

We therefore recommend that you open your favourite spreadsheet (probably Microsoft Excel, Apple Numbers or Google Sheets) and use that to follow along with the examples.

3.2 Using Assumptions

Assumptions are a significant aspect of operations management. There are just too may variables to consider, most of which are unknown or are difficult to predict. To perform analysis we have to simplify a situation so that we can make decisions and take action. It is common to face problems in industrial settings where the complexity of the situation takes over and appears to obscure any potential solution.

We manage this by assuming that certain things will behave in a predictable way or by removing detail from the problem. In theory this might reduce the accuracy of an answer (though this is not always the case), but in the real-world this might be the difference between taking action or being paralysed and not knowing what to do.

It is often better to try something based on some assumptions, and if it does not work as intended, we can use that new knowledge to develop another approach to tackling the problem. This iterative approach, if performed with some rigour, is a classic engineering approach to solving problems when there is no immediate solution.

However, it is important that our assumptions have some foundation to them; taking wild guesses as to how we should solve a problem is an unreliable method in the long run. We need to think about what we already know about a system, what we do not know, and what general principles there might be that can be applied to the problem under scrutiny.

What we already know about a system can be a challenging topic in itself, especially if there are a number of different people involved with that system. Each stakeholder will have a personal perspective that can enlighten or confuse this enquiry.

? Reflection question

Think of a time when you have been faced with an unfathomable problem. What did you find difficult to work out? What information would have made your situation easier to deal with? Alternatively, was there too much information?

Thinking about what we know usually identifies things that we do not know. It is typical for stakeholders to realise that they can cite a particular behaviour of a system, but they do not have the data to hand to explain how that behaviour occurs. Or, the data might be captured somewhere but it cannot be accessed conveniently or quickly enough.

Finally, we have the potential application of general principles. A process owner will have specific experience and knowledge of how an industrial activity relates to other business processes.

What about more general behaviours in systems? How might these be useful when thinking about specific industrial processes?

For example, how do seasonal sales patterns affect the flow of production orders through a system?

In the same way that an engineer can refer to physical properties when designing a product, we can consider more general principles that apply to most systems. To do this successfully we need to understand some fundamental concepts of how materials and information can be processed through industrial systems.

3.3   Operations Concepts

We shall now explore a collection of concepts that are useful for thinking about operations management. These concepts inform the approaches used in the rest of this book and so it is advisable to familiarise yourself with the detail to get the most out of subsequent chapters.

There are two areas of primary interest to the operations manager. First, how things flow through a system, whether it be a physical product or a piece of information. Second, how we can understand and make decisions based upon how we introduce materials (or information) into a system, and what we do with those items when they are work in progress (WIP).

The examples that follow in this book apply both to physical and information products as we are concentrating on *processes* essentially; as you become more familiar with the measurements and data that we need to perform industrial analytics, you shall start to see opportunities for process improvements in all areas of your organisation.

3.3.1   Cycle Time

When we assess a process our first measure that helps us is that of *cycle time* [1]. This is the average amount of time that we spend processing an item. It is easy and quick to calculate. It includes *all* aspects of the time required to produce something. This includes any waiting time, or time for reprocessing because of errors, transfer time between different operations and any time spent waiting *within the system*.

$$\text{cycle time} = \frac{\text{production time}}{\text{number of units produced}} \tag{3.1}$$

Let us say that a factory produces 47 items on average during an 8 h shift. Putting these numbers into the equation gives the following:

$$\text{cycle time} = \frac{8 \text{ hours}}{47 \text{ items produced}} = 0.17 \text{ hour} = 10.2 \text{ min}$$

A general objective is to reduce the cycle time as much as possible as this enables greater throughput. Of course, reducing cycle time is not always easy or possible.

A process supervisor uses cycle time as one measure of monitoring the performance of a process. As the demand for items varies, the process can be altered to keep the system stable. It is also the basis of forecasting. Frequently, process owners are asked to estimate the impact of a potential customer order. Having an understanding of the cycle time can assist such predictions.

? Reflection question

What effect does the amount of work in progress (WIP) in the system have on the cycle time?

An alternative approach to calculating cycle time is to use Little's law [2], which relates WIP, cycle time and throughput to each other:

$$\text{cycle time} = \frac{\text{WIP}}{\text{throughput}} \tag{3.2}$$

Since a good deal of the decisions that are made by a process supervisor needs to be made quickly, it is attractive to use Little's law as it does not require much in the way of measurement or analysis. The supervisor will intuitively know the average rate at which items are produced, which gives an approximate *throughput*.

WIP usually requires some counting—whether the jobs are physical or information—and this can be done swiftly. Together, these two values can lead to an approximation of cycle time, which is often deemed good enough for general decision-making.

Calculations that are simple tend to have limitations, and Little's law does not hold so well when there is a large degree of variation between the item types that are being processed. Similarly, different order quantities can distort the calculation by imposing an average value that may not be of any use in a system that has a lot of variation in demand.

Some composite processes are tightly coupled, and therefore the opportunity for variation between processing stages is minimal. In these cases, the calculation has little to offer.

3.3.2 Lead Time

In contrast to cycle time, lead time is the time from a request being made (an order) to the request being satisfied (receipt of the order). A retail example could be a customer ordering furniture from a shop. The retailer would inform the customer that the furniture would be delivered on a specific date in the future.

The time between ordering and receipt is the lead time for that particular order. Cycle time is a portion of the overall lead time; it is the time spent within the manufacturing facility. If we add the order processing time and the delivery time, we arrive at the lead time.

$$\text{lead time} = \text{date order received} - \text{order date} \tag{3.3}$$

For a customer who orders furniture on 1 February, receiving it on the 4 March:

$$\text{lead time} = \text{4th March 2021} - \text{1st February 2021} = 31 \text{ days}$$

Knowing the lead time is important for sales functions or customer-facing personnel in general as it enables reasonable forecasts to be made. Lead time is also used as a means of expressing what the customer experiences and can be a motivational tool for the factory to reduce cycle times.

3.3.3 Takt Time

Takt time is a measure that provides a production rate. It is used to give a budget of time for a given demand and can be used to check the feasibility of a potential order, or it can be used to identify either production inefficiencies or unexpected improvements.

$$\text{takt time} = \frac{\text{production time available}}{\text{number of items required}} \tag{3.4}$$

If a factory has an order for 80 items and it has a working week of 40h to produce those items, the takt time per unit 'budget' is calculated as follows:

$$\text{takt time} = \frac{40 \text{ hours}}{80 \text{ items}} = 0.5 \text{ hours per item}$$

If the supervisor knows that the cycle time of an item is typically greater than 30 mins, then more resources or time (later delivery date) shall have to be called upon to fulfil the order.

3.4 Using Concepts to Understand Systems

Knowledge of basic mathematical relationships between the concepts helps the supervisor make operational decisions.

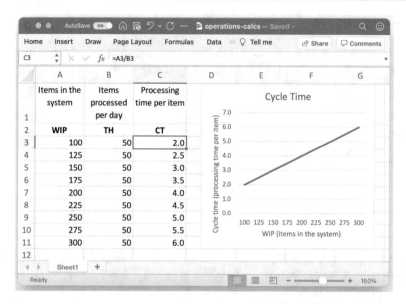

Fig. 3.1 Observing the relationship between WIP and cycle time. If a given throughput rate is maintained (TH), as WIP increases, cycle time (CT) extends also

By manipulating Little's law, we can focus on seeing what the effects of changing WIP volume within a system. This is done quickly with a simple spreadsheet as shown in Fig. 3.1.

In column A we have a set of values of WIP that increase from 100 to 300. Column B is the throughput (TH) rate, which is the number of items that are processed per day on average. Column C calculates the cycle time (the formula in MS Excel is '= A3/B3'). The line graph shows that as the WIP increases, so does the cycle time, which is undesirable as it will lead to late completion of the orders.

We can change the throughput rate to see what happens if we increase that. Figure 3.2 illustrates that the increase in cycle time is reduced, but it would still would impact upon delivery times.

If we reverse the calculation, we know that the desired CT is 2 and that we divide the WIP by TH to obtain CT.

So, if we divide WIP by CT, we can obtain the throughput target that needs to be achieved. Figure 3.3 illustrates this.

? Reflection question

- Take a business process that you are familiar with. What might be the effects of increasing the number of jobs (WIP) that are processed at any one time?
- Are these effects desirable or tolerable?

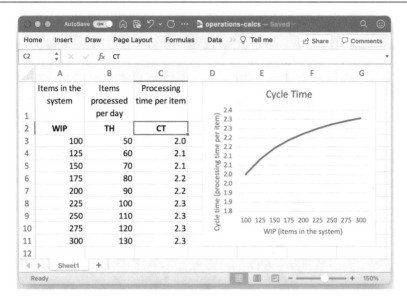

Fig. 3.2 Changing the throughput (TH) rate from a constant value to investigate what the effects of increasing it are in relation to cycle time (CT)

- What would the effect be of increasing the throughput rate (TH) and decreasing WIP for your business process?
- Under what circumstances might this cause problems?

3.5   Resource Utilisation

It is generally understood that high utilisation of a resource is preferable to low utilisation. The resource required to enable a business process has probably received some investment, so it is logical to assume that we should strive to utilise the resource as much as possible.

We shall now consider a single stage process that takes an input, does some processing on that input and then produces an output. Figure 3.4 shows the arrangement, which contains the following:

- inputs that enter the system at an arrival rate, which has a mean μ_a and standard deviation σ_a;
- an expected waiting time E(W);
- a processing time with a mean μ_s and standard deviation σ_s.

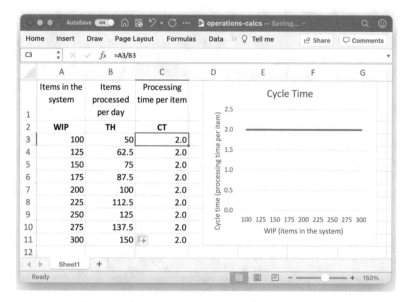

Fig. 3.3 Using the cycle time (CT) of 2.0 as a target, we calculate the throughput (TH) rate that is required for a given level of WIP

In addition to the above, we shall also consider

- Resource utilisation is the time that the process node is working, represented by ρ. This is a proportion and usually expressed as a percentage. We calculate utilisation by dividing the mean processing time μ_s by the mean arrival time μ_a.
- While we have a mean arrival rate (μ_a) all natural systems exhibit variability. The variability is quantified by the standard deviation, σ_a. We can thus calculate a coefficient of variation by dividing the standard deviation of the arrivals σ_a by the mean arrival rate, μ_a.
- Similarly we can calculate a coefficient of variation for the processing time by dividing the standard deviation of the process time σ_s by the mean process time μ_s.

Kingman's Equation (3.5) takes these values to provide an approximation of the expected waiting time for a given item from a system.

$$E(W) = \left(\frac{\rho}{1-\rho}\right) \cdot \left(\frac{C_a^2 + C_s^2}{2}\right) \cdot \mu_s \qquad (3.5)$$

We are able to explore the effect upon waiting time (the amount of time that an item has to wait to be processed) as utilisation is increased, by passing a set of different values to the equation.

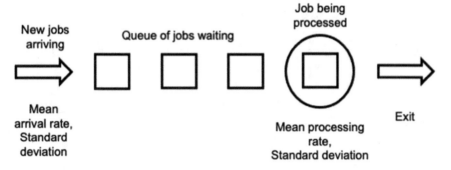

Fig. 3.4   Expected waiting E(W) time for an item increases as the process utilisation increases

As before, the simplest approach is to create a spreadsheet as shown in Fig. 3.4. Column A contains a range of values for p, which is resource utilisation, from 0.50 to 0.95 (50–95%).

Column C calculates

$$\left(\frac{\rho}{1 - \rho} \right)$$

Column F calculates

$$\left(\frac{C_a^2 + C_s^2}{2} \right)$$

For our simple system, we observe that there is a direct relationship between resource utilisation and waiting time, such that an increase in one also inflates the other (Fig. 3.5).

Our use of basic mathematics illustrates two important points for process managers. First, if we attempt to push more jobs into a system, the average cycle time will increase. A common reaction to an increase in WIP is to force the utilisation higher.

But our second finding illustrates that if we increase resource utilisation, the overall waiting time goes up, increasing the cycle time and eventually resulting in late deliveries.

Thus, WIP and resource utilisation are important measures, but they need to be managed collectively.

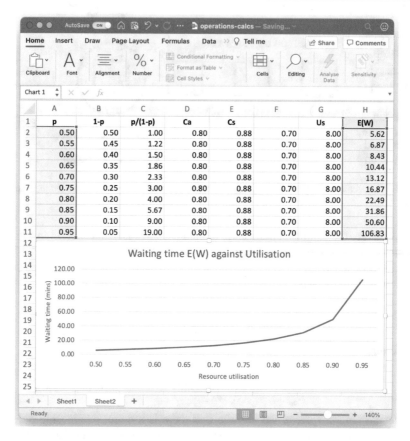

Fig. 3.5 Expected waiting E(W) time for an item increases as the process utilisation increases

3.6   Learning Activities

This chapter has introduced both some concepts, plus some basic approaches to analysing systems. Central to the use of these techniques is data. What data do you have access to?

A good exercise is to create a list of the sources of data that exist within your organisation and then categorise them by the source. Some examples might be spreadsheets; exported data from enterprise resource planning systems; barcode scanners; plant data, etc.

? Extension exercise

Most businesses contain a collection of processes, some of which are interconnected. There are also more extended *supply chains* where processes between organisations are interlinked. Using what you have learned from this chapter:

1. Make a list of the potential interruptions to process flow.
2. Consider what changes you could propose that will help reduce the impact of these interruptions. Think of the data that you will need, and how you will monitor the system's performance.

References

1. Hopp WJ, Spearman ML (2001) Factory Physics: Foundations of Manufacturing Management. Irwin/McGraw-Hill, Boston
2. Little JDC (1961) A Proof for the Queuing Formula: $L = \lambda W$. Oper. Res. 9(3):383–387. https://doi.org/10.1287/opre.9.3.383

Data for Production Planning and Control

4

4.1 Historical Attitudes Towards the Use of Data

The development of current attitudes towards planning and control methodologies, and their implied data requirements has followed the rise in *mass production*, initially moving from home working into factory working, and can be said to have its (current) origins in scientific management [5].

Scientific management originated by Taylor [9], operated through the (prior) analysis of the stages within the manufacturing process and commonly termed time and motion studies, measuring/obtaining data about the manufacturing system (using the available historic data). But this approach, using static data, did not allow for the (sometimes) chaotic nature of the production process.

This, using static data, approach has worked, obviously, at Taylors firm but more noticeably at the Ford Motor Company [10] in the 1920s when efficient operation of production lines was planned and controlled (timed) using measured historic data, to enable the mass production of cars minimising the need for current data.

It must be observed that at the Rouge River Ford plant in the 1920s, before computers, the production of only a single product (Model T Ford) reduced the data requirement and reduced the need for accurate forecasts for each individual car. Similarly at Toyota in the 1950/1960s although computers were available they were not used for (manufacturing) data collection; consequently they developed the Kanban control system to enable production control and forecast job completion times.

However, it has been shown that such methodologies, using limited data collection, an ad hoc approach to allocating production time, and resource-intensive progress chasing often have resulted in unsatisfactory planning and control outcomes.

A typical example of such an ad hoc approach is a company producing many products for sale (mainly) through a chain of retail outlets, each outlet reporting (by post) its weekly sales (by product) to the company.

Thus, although this company had much data it was unable to process this data to produce meaningful forecasts (to enable factory planning), so its first step was to

© Springer Nature Switzerland AG 2021

R. Hill and S. Berry, *Guide to Industrial Analytics*, Texts in Computer Science,
https://doi.org/10.1007/978-3-030-79104-9_4

aggregate this data to give a total sales per item; then the forecast sales for the next year were calculated through the addition of a fixed percentage to these aggregated values. This policy leads to problems of oversupply of some items at some outlets, where demand was falling, and a lack of supply of other items at the same outlet, where demand was rising.

A firm will consider a planning and control system to be good if it ensures that the majority of items produced are of suitable quality and delivered on time (at a minimum cost) and for those not on time the size of the delay will be known.

> Failure to schedule effectively results in customer dissatisfaction and subsequent lost orders, and is thus a key dimension in the success or failure of small firms (Holliday [4])

Effective scheduling is dependent upon the existence and use of reliable data about the status of the firms production and sales functions.

4.2 Need for Data Within the Production Area

Planning and control within manufacturing is dependent upon the availability and use of timely quality data.

Zapfel and Missbauer [11] have indicated that much of the time spent by a job within production areas is waiting time, thus indicating the need for data describing the status of each job as it progresses through the factory, to enable the firm to determine the expected job completion times.

Historically a problem (for a firm) was the collection of the required quantity and quality data about all jobs within the system (in manufacture or queueing) on time to be able to implement good production control systems, a major problem being the cost of collection and analysis of the selected data.

However with the advent of computer-controlled systems the cost of data collection has been (much) reduced, and the problem for the firm has been transformed from the position where data collection was costly and had to be carefully considered into a state where their problem is the selection of the most appropriate data, from the large quantity of computer generated data, to enable effective and efficient planning and control.

Transforming the company's problem from:

- What data, and how much, to collect to
- What data to use, how little, from the quantity of computer generated data.

This problem, of what data to collect, is best illustrated through the widely varying importance attached to stock levels within a just-in-time environment.

Schonberger [8] suggested that minimal stock levels are very important in a JIT environment, thus suggesting that very good quality data will be required about stock level, while Brown [1] questioned whether stock levels are really so important in a JIT environment, suggesting that some stock data is necessary, and Ohno [6] says that stock levels are not relevant with a JIT system, suggesting that stock data is less important.

The objective is being to enable effective planning and control within the firm to satisfy the objectives:

> Production Control provides the foundation on which most industrial controls are based. To a very large extent controls used in industry only watch the side effects produced by changes in the flow of materials, and without efficient production control it is generally impossible to have effective control of any sort (Burbidge [2]).

Consequently the formal requirements for a (planning and) control system have been identified [7] as:

- Objectives must exist for the activity being controlled.
- The activities outputs must be measurable using relevant criteria or indicators.
- A predictive model of the activity being controlled is required.
- There must be capacity for taking remedial action where target delivery dates may not be achieved.

All these formal requirements indicate the need for sufficient timely data to be able to set delivery dates and monitor and control the system so that these dates could be achieved.

4.3 Planning Problems Resulting from Lack of Appropriate Data

These occurred when a manufacturing firm operated either

- without any planning;
- no real planning or
- planning with no recognition of the state of the system.

All such firms are being identified by their lack of use of data concerning the status, work in progress/status of their manufacturing facilities. Consequently the operation of the firm will fall into one of the two ineffective states as shown in Fig. 4.1: Both ineffective states lead to the use of a greedy approach to production planning but without any consideration of the quantity of work in progress, and as a consequence the firm will (often) be uncertain about the probable job completion date for a newly received order.

State 1

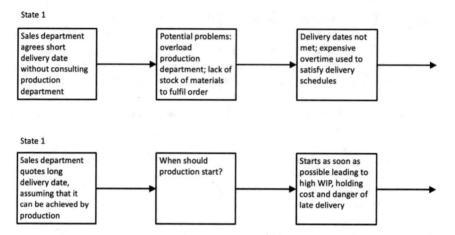

State 1

Fig. 4.1 Common effects upon production caused by a lack of planning data

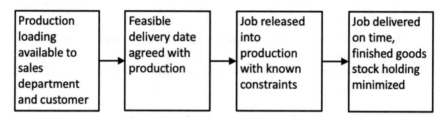

Fig. 4.2 Result from planning with appropriate data

To overcome this problem these firms often quote a long delivery lead time (much longer than the required production time) to ensure that delivery can be on time but at the expense of high stock holding costs and excessive work in progress.

4.4 Planning with Appropriate Data

These inefficient states can be compared with the state of those firms where the company used appropriate data to plan and control its production systems. Here data concerning the status of the workshop enables sales to quote satisfiable data to the customer while reducing the quantity of finished stock held in a warehouse (reducing stock holding costs) and enabling an efficient use of the production facilities (Fig. 4.2).

4.5   Need for Optimality in Production Control and Scheduling

A firm will need to have both an appropriate production planning and control system and appropriate set of data to enable the efficient and effective use of their production facilities. Thus, a firm will wish to construct/develop and operate a production planning control system and define the dataset necessary so that their operations will be optimised.

However optimality could be described or measured in term of one or more of the criteria as follows:

- efficient usage of the production system, reduce waiting time;
- maximum usage of the production system, minimise machine idle time;
- use a minimum number of production units to satisfy demand;
- always supplying customers on time;
- minimise the quantity of work in progress;
- minimise stock holding costs;
- maximise the contribution to profit;
- maximise the number of new orders (sales) while delivering on time.

To be able to satisfy any of these objectives there will be a need for the firm to develop certain operational characteristics:

- ability to forecast job completion times;
- ability to agree delivery times with customer;
- Agree delivery times and make effective use of Greedy or JIT planning;
- efficient and effective use of resources, ability to know the system status;
- satisfying customer demand on time.

Each definition of optimality implies that the information present within the organisation needs to be shared between the various functions within the company leading to information flows shown in Fig. 4.3, and each operational characteristic will require some data concerning the status of all current jobs at that time of arrival of a new job or set of jobs. Thus, an appropriate system will ensure that sufficient (but not excessive) stocks of raw materials and subcomponents are available for production, whilst also providing management information that enables realistic delivery time estimations for the customer.

The best system will therefore need to optimise the system 'costs' where the system cost will be constructed from

- cost of early completion;
- cost of late completion;
- raw materials and subcomponent storage holding, investment and shortage costs;
- operational costs.

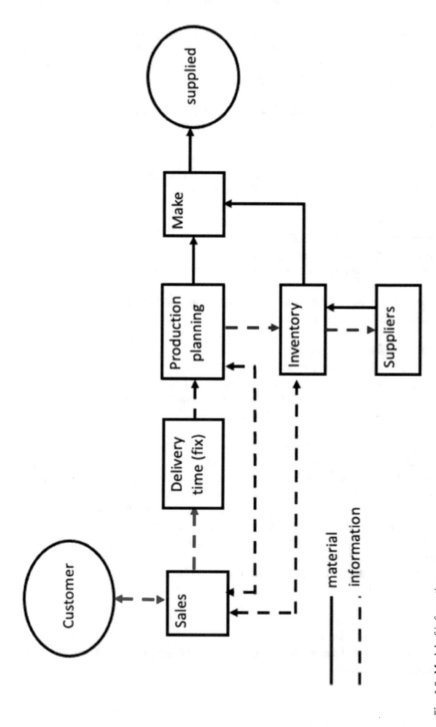

Fig. 4.3 Model of information

Table 4.1 Comparing stock holding of Greedy versus Just-In-Time production planning approaches

Greedy production planning				
Job	Duration	Due date	Available	Holding
1	5	18	5	13
2	3	15	8	7
3	8	20	17	3
Total				23
Just-In-Time production planning				
Job	Duration	Due date	Available	Holding
1	5	18	9	9
2	3	15	12	3
3	8	20	20	0
Total				12

The operation of any production planning and control system will require data about the system. The simplest dataset uses the number of jobs within the production system, the current order book, to estimate job completion times this approach implying the use of a regression model.

A major problem with this approach is the quantity of data needed to establish and operate a reliable *regression modelling*-based forecasting model.

Having agreed a feasible delivery time with the customer (planning) the next stage is for the production planner to update/reschedule the existing and new jobs to produce a new (detailed, machine loading) production schedule (scheduling).

Effective production planning and control depends upon the ability of the firm to

- *Plan*—set/agree delivery times;
- *Schedule*—release job into production;
- *Control*—ensure that the desired completion/delivery times will be achieved.

For scheduling, the two extreme alternative policies are Greedy and lean production planning systems Fig. 4.5.

To compare the use of these alternative policies consider an example where there are (currently) three jobs to be scheduled with due dates [9], respectively, Table 4.1 lists the job data and indicates the total stock holding resultant from the use of each approach, and Fig. 4.4 shows the two schedules of work.

1	2	3	4	5	6	7	8	9	10	11	12	13	14	15	16	17	18	19	20	21	22	23	24
GREEDY SCHEDULE																							
Just in Time Schedule																							

Fig. 4.4 Comparing Greedy and JIT schedules

Comparison of the Schedules:

- *Greedy Schedule*—has higher stock holding costs;
- *Just-in-Time Schedule*—more likely to need control actions following delays in production.

Now consider the effect on these schedules when the sales department wishes to introduce a new job into the system; their first step being to determine a delivery date. Fig. 4.5 shows the feasibility of the existing (greedy) schedule. At the arrival of a new job into the system we investigate possible delivery dates: Try Day 19, then Day 22 (Fig. 4.6).

Adding the new job gives the new loading plot, Fig. 4.7, indicating that this date is not feasible but does indicate that there is a feasible delivery date.

Day 22, this plot (Fig. 4.8) indicates that this, or a later date, will be a feasible delivery date.

This graphical approach to production scheduling is described in more detail in Appendix B.

The same conclusion, first possible delivery date (Day 22), would also be derived starting from the JIT schedule.

- *Greedy Scheduling*—No rescheduling needed.
- *Just-in-Time Scheduling*—All jobs need to be rescheduled.

The advantages of these alternative approaches can be summarised by

- *Minimising holding costs*—JIT;
- *Minimising control*—Greedy;
- *Reducing rescheduling*—Greedy.

In general for high value products adopt a just-in-time approach, and for low value delivery-sensitive items adopt a greedy approach.

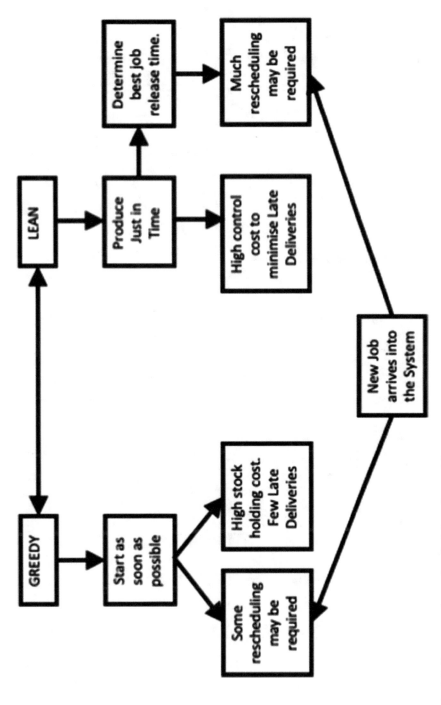

Fig. 4.5 Two approaches to the scheduling of a new job

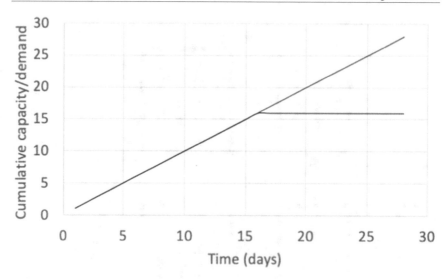

Fig. 4.6 Comparing schedule with capacity

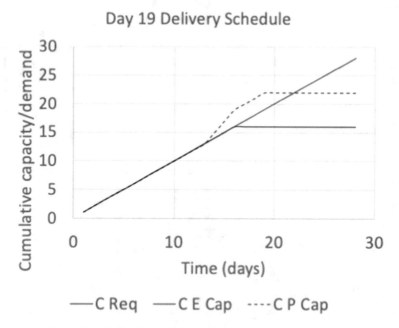

Fig. 4.7 Machine loading plot for Day 19 of a schedule

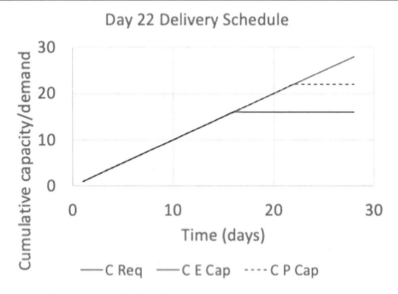

Fig. 4.8 Machine loading plot for Day 22 of a schedule

4.6   Deriving Generic Models for Planning and Control

To be able to determine an appropriate quantity and quality of data to collect, or access, to enable the efficient and effective use of a firms' production facilities it is first essential that the planning and control systems commonly used within such small manufacturing firms are identified and generic planning and control systems synthesised from this information.

The analysis of these commonly used approaches to production planning and control, often classified as push- or pull-based systems, leads to the derivation of a generic planning and control model, hence indicating the data required to enable effective production planning and control.

Production planning and control systems have been described as being based on either a 'pull' or a 'push' philosophy. However, Buzacott and Shanthikumar [3] claim that pull and push systems are special cases of a more general production planning and control algorithm. The objective here is the synthesis of a general model, and from this model deduce the information requirements of such systems.

The model developed by Buzacott and Shanthikumar uses the notation:

- z_j—inventory at stage j;
- k_j—maximum number of jobs allowed at stage j;
- r_j—delay between receipt and transmission of order at stage j;
- c_j—number of processors at stage j.

Table 4.2 Formal description of planning and control methods

		Allowed stock z_j	Limit on jobs k_j	Information delay r_j
PUSH-based systems	Produce to order	0	∞	1
	Base stock system	0	∞	≥ 1
	MRP II	≥ 0	∞	≥ 1
PULL-based systems	Kanban	>0	z_j	≥ 1
	Local control	$>c_j$	c_j	1
	Interval control	≥ 0	$z_j + k_{j+1}$	1
	CONWIP	0	Z	1
	OPT	≥ 0	Factory wide control (F)	≥ 1

Table 4.3 Generic planning and control paradigm

System type	z_j	k_j	r_j
PUSH	≥ 0	∞	≥ 1
PULL	≥ 0	K	≥ 1

The commonly named planning and control systems (approaches) can be considered to fall into one of two categories:

1. MRP push-based systems: produce to order; base stock system; MRP II;
2. JIT pull-based systems: Kanban local control; CONWIP interval control; optimised production technology (OPT); theory of constraints (TOC).

These approaches can be summarised, using this notation see Table 4.2, with the techniques grouped using a 'push–pull' division. By simplifying the relationships, this information can be used to demonstrate that each of these methods has been derived from the generic production planning and control paradigm given in Table 4.3.

Developing Buzacott's analysis most differences in Table 4.4 is concerned with the value k_j 'the maximum number of jobs allowed at stage j', either limited or unlimited capacity.

The relationship between the different approaches becomes apparent, from this generic model, indicating a focus on either planning or control. For example, the first group of methods can be considered to have been derived from the MRP paradigm—a *push* model, leading to the generic model representing push systems as Table 4.5. Thus, it follows that a *produce to order* system, which is commonly used in small jobbing/manufacturing firms, and *base stock* systems are special cases of a generic

Table 4.4 Push model

System type	z_j	k_j	r_j
PUSH	≥ 0	∞	≥ 1

Table 4.5 Push systems

Generic	Special case
MRP	Produce to order; Base stock system

Table 4.6 Pull model

	z_j	k_j	r_j
Pull model	>0	L	≥ 1

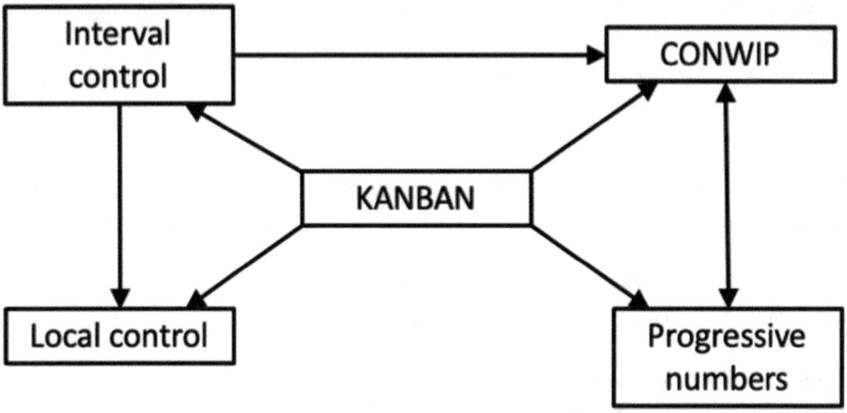

Fig. 4.9 Pull model

MRP system suggesting that the data needs of such firms are a 'subset' of the data requirements of an MRP system.

Similarly, the second set of planning and control systems will have many features in common, suggesting the *pull* paradigm.

The generic pull model is given in Table 4.6, in which L represents the limit on k_j in all these approaches. The data requirement of pull systems is concerned with the value k_j and its maximum allowed value L.

Relationships between these approaches are illustrated in Fig. 4.9. This analysis can now be extended to develop the system selection model of Fig. 4.7, and this model indicates how firm size and product type influence the firms chosen approach to planning and control suggesting the data requirements from each approach.

This model suggests that in a 'push'-based environment order status, knowledge of the number of jobs present is an important factor while in a 'pull'-based environment material flow is the important factor. However as any planning and control system will combine elements from both these paradigms then it follows that the important

Table 4.7 System selection model

Strategic variables		Shop floor control approach	
		Push-type	Pull-type
Product	Type	Special	Standard
	Range	Wide	Narrow
Individual product volume per period		Low	High
Versatility	Product mix	High	Low
Delivery	Speed (achieved by)	Schedule change	Finished goods stock
	Schedule changes	More difficult	Less difficult
Process choice		Jobbing	High volume
		Low volume batch	Batch/line
Manufacturing Control	Key feature	Order status	Material flow
	Basis	Person/system	System
	Ease of task	Complex	Easy

data, for planning and control, will be concerned with the number of jobs present and material consumption, which will also be dependent upon the number of jobs present.

This analysis indicating that to enable efficient and effective production planning and control data should be collected concerning

- number of jobs within the system or
- number of jobs at each stage within the system.

4.7   Production Planning in Manufacturing: Small Case Study Results

To be able to validate these proposed generic models a set of case studies developed around a selection of manufacturing firms. These models provided detailed pictures of the planning and control systems typically used in such firms and hence enabled the identification of the data requirements of such firms.

The case study firms ranged in size from micro- to medium-sized firms each producing goods to satisfy customer requirements (orders).

These case studies provided information about

- *The production system*: number of stages; number of processors at each stage; number of production types;
- *Planning and control*: planning system used; control system used;
- *Demand patterns*.

Table 4.8 Formal description of case study

	z_j	k_j	r_j
Pull model	0	>0	1

It must be noted that all the case study firms operated as Flow Shops, similar to a production line, and all firms had, pragmatically, developed efficient approaches to production planning and control systems.

Furthermore the planning and control methods designed and used by these firms can, best, be described as 'pull' systems allied with a CONWIP approach to controlling the quantity of work present in the workshop.

Typically the firms experienced a fluctuating seasonal demand pattern with orders arriving at random and can be for single or multiple items of one or several product types.

These case studies were used to identify the information necessary to develop the simulation models which were used to determine the data needs to enable efficient and effective production planning and control.

4.8   Planning and Control in the Case Study Firms

An analysis of the information from these small manufacturing firms leads to a formal description of the planning and control systems. Important features of their operations were as follows:

- These firms manufacture goods to order and hence hold little, or no, stocks of finished goods or work in progress.
- The maximum number at each stage is determined by the manager.
- Jobs tend to move between stages individually.
- No delay on order release.

The formal description of these manufacturing firms is provided in Table 4.8. The number of stages and number of processors at each stage are detailed in Table 4.9. Additionally for all firms the dominant stage, that stage with the lowest capacity, was located at one of the later stages in the production process and for most firms at the final stage.

It therefore follows that the data for these firms needs will be

- number of jobs within the system (primary);
- status of each job in the system, location and status at the location (secondary).

Table 4.9 Production stages and processors for each facility

Firm	Number of stages	Processors at each stage
A	2	1 at each stage
B	3	1 at each stage
C	3	N_i at stage i, $N_1=1$
D	3	N_i at stage i, $N_1=1$
E	3	N_i at stage i, $N_1=1$
F	4	1 at each stage
G	2	N_i at stage i
H	4	N_i at stage i
J	6	N_i at stage i

Currently this level of data can be available directly from these production stages in the form of Internet data. Simulation models have been constructed to be able to determine the minimal information set required to enable the implementation of efficient and effective planning and control systems within a manufacturing firm.

4.9   Manufacturing Production Systems in Case Study Firms

A manufacturing workshop can be considered to consist of a series of queues and servers typically, as indicated by the models constructed from the small case studies, and a small firm will tend to have a production system consisting of three stages as follows:

- Stage 1—Prepare;
- Stage 2—Make;
- Stage 3—Finish.

where jobs progress sequentially through the system, and jobs may join a queue of waiting jobs before each stage, Fig. 4.10.

A larger firm may have the same three stages but with more like processors at each stage, or more stages or more stages with several like processors at each stage. Figure 4.11 illustrates a system where the new job (at each stage) accesses the first available processor, and Fig. 4.12 illustrates a system where a large workshop is subdivided into a set of production lines.

Note that: A large firm may have several stages within their production system with many processors at each stage, not all processors being the same.

As a firm grows, from a being a small firm with a single processor at each stage, the production system could become like those shown in Figs. 4.11 and 4.12 where there are different numbers of processors at each stage, and each job progresses onto the first free machine or onto its most preferred processor, with one stage being dominant (having the lowest capacity per time unit).

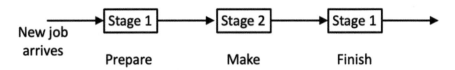

Fig. 4.10 Small firm production schematic

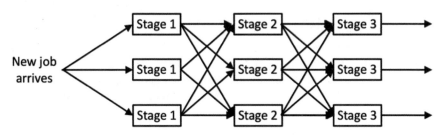

Fig. 4.11 Three processors at each stage

4.10 Summary

The information gained from these investigations into manufacturing processes (number of stages and number of processors at each stage) was to be used to enable the construction of models to enable the determination of the data needs (quantity and quality) to enable effective planning and control within such firms. The results indicated that the information needs were concerned with

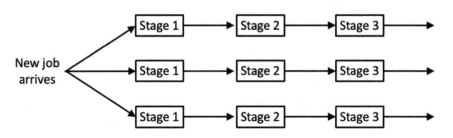

Fig. 4.12 Three production lines

- the number of jobs in the system;
- the location of the jobs in the system;
- the quantity of work remaining for the jobs in the system.

The models were used to enable the simulation of

- small firms;
- firms growing from a small firm into a larger firm and
- large firms.

where the production system, number of stages, remains the same as the firm grows. In the context of such manufacturing firms effective planning and control will be measured through the firms' ability to set, or agree, 'good' delivery dates, where a good delivery date will be such that the firm can (reasonably) expect to be able to deliver the required goods on time to the customer, and efficiency is measured in the quantity and quality of date required to be able to derive these 'good' delivery dates.

These simulations will also enable an investigation into the effect of the dominant stage (by processing time) in the production system on the firm forecasting (data requirement and process), evaluating the effectiveness and efficiency of the derived methodologies and indicating appropriate approaches to planning and control for such firms.

4.11   Learning Activities

? Exercise

1. How did the Ford Motor Company Rouge River Plant reduce the need for 'information' within the firm and hence simplify their production planning and control systems?
2. How did the Ford Motor Company's information systems at their Rouge River Plant influence the development of production planning and control systems?

? Extension exercise

1. Discuss why growth within a firm can cause the need for change and refinement in its approach to production planning and control. Make reference to the various characteristics of a business that can affect its capability to plan and execute production schedules.

2. Compare and contrast the advantages of just-in-time and Greedy approaches to production planning and control with reference to the need for production data. Use examples to explain your answer.

References

1. Brown KA, Mitchell TR (1991) A comparison of just-in-time and batch manufacturing: the role of performance obstacles. Acad Manag J 34(4)
2. Burbridge JL (1978) The principles of production control. McDonald and Evans
3. Buzacott JA, Shanthikumar JG (1993) Stochastic models of manufacturing systems. Prentice Hall
4. Holliday R (1995) Investigating small firms. Routledge
5. Hopp WJ, Spearman ML (1996) Factory physics: foundations of manufacturing management. Irwin
6. Ohno T, Mito S (1988) Just-in-time. Productivity Press
7. Otley DT, Berry J (1980) Control organisations and accounting. Account Organis Soc 5(2)
8. Schonberger RJ, Knod EM (1994) Operations management: continuous improvement. Irwin
9. Taylor FW (1911) Scientific management. Harper Brothers
10. Wilson JM (1995) Henry Ford's just in time system. Int J Oper Prod Manag
11. Zapfel G, Missbauer H (1993) New concepts for production planning and control. Euro J Oper Res 67:297–320

Part II
Methods

Using a series of worked examples and in-depth reflection questions and learning activities, the use of analytics techniques is demonstrated.

Simulating Industrial Processes

5

5.1 Understanding Business Operations

Manufacturing facilities vary in both size and complexity. One factory might have two or three areas where different processes take place. Another factory might contain hundreds of machines. Each of the factories will have evolved to cope with the manufacture of different products, different mixes of order types, different customer demand profiles, varying quality of raw materials, unpredictable machine breakdowns and so on. The list of possible interruptions to a neatly ordered continuous flow of efficient output is endless.

The shop floor supervisor manages these variations using their analytical skills, as well as the experience that they have accumulated while working with such systems.

At some point during the working week, they'll be required to answer the following questions:

1. When will order X be finished?
2. How much stock is tied-up in the factory?
3. What is the utilisation of the work centre?

These questions might be asked by different stakeholders. Question 1 probably comes from the customer, via the sales department, perhaps because the order is late.

Question 2 might come from purchasing who are concerned with reorder quantities for input materials. Or it might be the accountants who are assessing cashflow.

Question 3 is definitely asked by the accountants, so that they can put a measure on the production potential of a factory. But it is also posed by the production planners who want to find additional production capacity for more customer orders.

In a smaller organisation most of these roles may be undertaken by the same person. In larger companies the functions will be separate departments. Whatever the size of the facility, the questions are the same. The answers are likely to be the same also.

© Springer Nature Switzerland AG 2021
R. Hill and S. Berry, *Guide to Industrial Analytics*, Texts in Computer Science,
https://doi.org/10.1007/978-3-030-79104-9_5

When faced with such questions there are too many variables to consider to make a reasoned judgement. Such answers start with 'it depends'.

Attempting to quantify the lateness of an order is dependent upon the jobs in front of the late order, the reliability of the process, whether the operator is working at peak performance, the quality of the tooling and raw materials, etc. If the process in question is fed, or feeds into other processes, the opportunities for error are compounded. This leads to the use of estimates which are generous and therefore may build inefficiencies into how we manage the overall operations.

What we need is a model of the facility. This model captures the essential characteristics of the manufacturing unit and let us change some of those characteristics so that we can see what the effects might be.

Our supervisor might have had an idea to reduce the batch sizes of their orders, but not felt able to try it out as their machine utilisation measures might drop. Or if something went wrong and an order was late, the change initiated by the supervisor might be cited as the cause of the reduction in output.

But if that change could be applied to a model, that has no physical connection to the real facility, perhaps we could learn more about how the system behaves. If we understand the system better, we stand to make better decisions in the future.

This practice is referred to as simulation, and it has long been the preserve of industrial mathematicians, or scientists who study operational research. Such work creates a lot of value for organisations, by creating models and allowing production personnel to experiment with different strategies.

However, these mathematical approaches are often inaccessible and significant training is require to interpret the models. But we can often obtain much of the benefit of modelling without the need for advanced mathematical skills, and this is the approach that we adopt in this book.

In the remainder of this chapter we are first going to look at an everyday approach to creating a model of an industrial process or service. We shall consider how we can ask questions of the model and use that to improve our understanding.

With this understanding we shall then look at building a simple tool to simulate the operation of the process/service. This simulation will produce results that we can use to experiment with different scenarios so that when we go back to the shop floor, we can take more informed actions.

5.2 Queues and Queueing

Let us assume that you visit the local supermarket to buy a few items. You select your items and make your way to the checkouts to pay for the shopping. There are a number of checkouts in the supermarket but for some reason only one of the checkouts is operational. You are not the only customer in the store, and there are three other people already at the checkout, waiting in a queue to be served. When they have been through the checkout, it will be your turn to be served. Figure 5.1 shows the scenario.

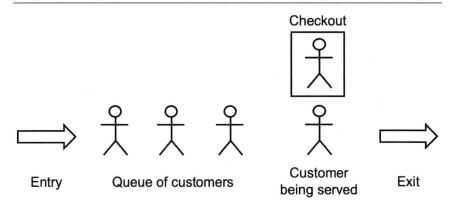

Fig. 5.1 Supermarket queueing with one checkout

We are going to assume that each person and their shopping in the queue represents one job.

Just for a moment, think about your answers to the following questions:

1. When will your job (you and your shopping) be finished?
2. How many jobs are there in the queue?
3. What is the utilisation of the checkout?

You may recognise these questions from earlier. What was your answer to Question 1?

Since we don't know how long it takes to process any of the shopping, we would have to say it 'depends'. It depends on how much shopping each person in the queue has; this might range from a hand basket to an over-laden trolley.

Question 2 is a little simpler. We know that each person and their shopping is classed as a job, so we just count the number of jobs in the queue. If there are three jobs waiting in front of you, there must be a job in progress at the checkout, which makes four jobs. And then there is you, bringing the total to five.

And what about Question 3?

When thinking about utilisation we need to consider potential interruptions such as:

- the checkout operator being changed at the end of a shift;
- a request form a customer services supervisor for a price because the barcode on an item is unreadable;
- a power cut causing the till to stop working.

If there is a queue of customers, and there are no disruptions to the actual process, we can assume that the checkout is kept busy. Once the queue becomes zero (all the jobs have been processed), or there is an interruption, the checkout becomes idle and the utilisation drops.

Now that we have a basic representation of our supermarket checkout in place, let us see how we can alter the performance of the system.

The supermarket manager realises that if customers have to queue for too long they may become frustrated, or even leave the store without making a purchase. This is not good for business, so another checkout is opened up (Fig. 5.2). Now, you approach the checkouts and find that there are two checkouts working. Each

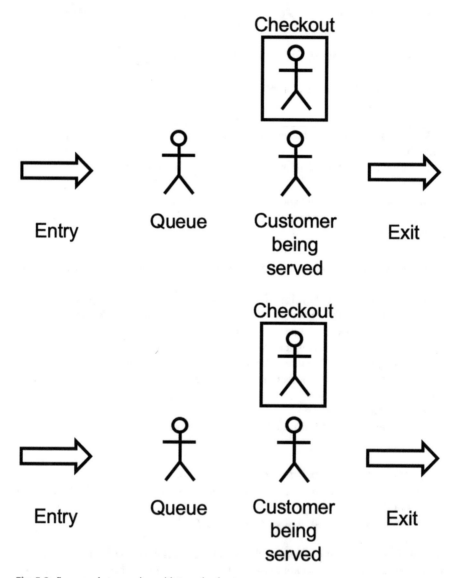

Fig. 5.2 Supermarket queueing with two checkouts

checkout is processing one job each, with a queue of one job waiting also. You are free to join either queue.

Let us assume that it takes the same amount of time to process each job. If that is the case, since either queue is shorter your job shall be processed faster. The utilisation of the checkouts reduces; however, unless there are more jobs arriving behind you.

We can thus deduce that there is some form of relationship between the number of available checkouts, the number of jobs to be processed, and the overall time taken to process an individual job.

If the supermarket manager had such a model, they could experiment with the optimum number of checkouts to service their customer demand patterns. This would then help them allocate the correct number of checkout staff for busy periods, while reducing the instances of checkouts being idle during quieter times.

The model would permit them to plan for seasonal adjustments in shoppers' behaviours. But, if the model can be executed quickly, it could also be a tool to explore a scenario that is unfolding—such as a large influx of customers that were unexpected—and this is where modelling and simulation can become a powerful tool for the management of business operations.

5.3   Modelling an Industrial Process

We shall now consider an industrial scenario. A joinery company produces wooden window frames. Each of the frames is cut from lengths of timber that are shaped and cut to length by a machine.

The company receives orders of varying quantities of windows, which means that varying numbers of timber lengths are required from the first machine. The company only cuts timber lengths for orders and does not make products to put into stock.

Each order is considered to be a job. Just as was the case with the shopper and their variable amount of shopping, each job can vary in size.

Each job must then spend a certain amount of time waiting in a queue, before being processed by the machine. The total time that the timber is in the system is the queueing time + processing time.

Both the queueing time and the processing time are dependent upon the size of the respective order.

We can see now that the model for creating lengths of timber window frame is exactly the same as our first supermarket model. We have jobs, a queue, and a processing station, where the actual work gets done. This scenario is shown in Fig. 5.3.

We can also start to see that even with a few numbers missing (the time taken to process a job, or the size of the shopping basket) that we can start to ask questions of the model and to infer potential relationships that might describe the eventual behaviour of the model.

For instance, what impact is a longer queue going to make on (a) resource utilisation, and (b) the overall time that a job spends in the system?

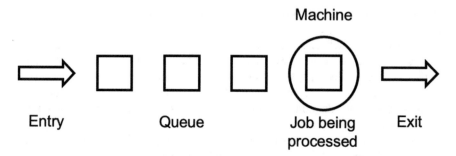

Fig. 5.3 Queueing model for a single industrial machine

A longer queue suggests that there will be less interruption to flow, so the utilisation will be higher. However, the longer the queue the more time that a particular job takes to be completed, so the delivery time is longer.

The next stage is to build a simulation so that we can verify our assertions.

5.4 Designing a Process Simulation

We now have an illustration of how we can model a single industrial process. That model is part of the initial specification of a simulation that we can execute. The simulation will execute a virtual production run, and that will give us an idea of how the model can perform.

The simulation allows us to change different parameters of the model, without incurring the cost or disruption of moving physical plant around.

So far, our model describes:

- a process of material conversion, where lengths of timber are given a profile and then cut into shorter lengths that are suitable for window frames;
- a single machine that performs the operations described above;
- each job is processed one at a time. Multiple jobs cannot be processed simultaneously;
- jobs arrive for processing and wait to be processed in a queue;
- a job that has been processed is deemed to be complete and exits the system.

We now need some more information to allow us to build the simulation. First, we should describe the rate at which jobs arrive for processing. Second, we need to specify the time taken to process a job.

Third, we need to consider whether there is any variation in the size of a job. For this first example we shall assume that each job requires roughly the same amount of time to process. We shall explore variable job sizes later.

There are many different simulation tools that can be used to build queueing models. We shall be using 'Ciw' (which is Welsh for 'queue').

Ciw is a simulation framework that uses Python and as such is free to acquire and use. For installation instructions, see Appendix B.

Ciw has three parameters that are of relevance to our industrial process model.

1. *arrival_distributions*: this is the rate at which jobs arrive to be processed. We shall assume that the jobs arrive approximately every 15 min, or four times per hour;
2. *service_distributions*: this is the time that each job spends being processed, or the time taken to do the shaping and cutting to length of the timber by the server (machine). We shall assume that each job takes 15 min;
3. *number_of_servers*: this represents the number of machines at a workstation. In our example, we have one machine, or one server.

It is important to note at this point the difference between parameters that are static, and those that might vary.

For instance, for a given simulation we can assume that the number of machines (servers) does not alter, so we give it the value of 1 as we want to investigate the scenario with one machine.

However, while we can say that jobs arrive at a rate of four times per hour, or every 15 min, that isn't strictly realistic.

Sometimes there are interruptions to the deliveries. A forklift truck might drop the timber when loading it from the lorry, or there may be a physical blockage preventing the wood being placed next to the machine.

Similarly, the time taken to process the timber won't always take 15 min. This is just an approximation that—on average—takes 15 min. Sometimes the timber might blunt the cutting blades of the machine, and it will take longer to finish the operation. When the tooling is freshly sharpened the machining time will be less than 15 min.

We want our simulation to take account of these variances, and we do this by specifying a distribution function. This tells the simulation to use a variety of values, whose mean is the arrival rate that we are suggesting.

So, for an arrival rate of 15 min, the simulation will generate a set of values that vary, but have a mean of 15 min.

This allows the simulation to be more realistic as it will take account of naturally occurring variations in waiting and processing times.

We are now ready to build the simulation.

5.5 Building the Simulation in Ciw

Create a new text file called timber_conversion.py We shall enter some snippets of code now to quickly create a simulation to produce some results. Try not to worry about some of the details just yet as they will be explained later. What is important is to execute a simulation so that we can start to understand the timber conversion process better. First, we specify the arrival and service distributions, along with the quantity of servers:

Python Code

```
import ciw

N = ciw.create_network(
    # jobs arrive every 10 minutes, or 6 times per hour
    arrival_distributions=[ciw.dists.Exponential(0.1)],
    # jobs take 15 minutes to process which is 4 jobs
    completed per hour
    service_distributions=[ciw.dists.Exponential(0.07)],
    # the number of machines available to do the processing
    number_of_servers=[1]
)
```

You might have noticed that the value contained in [ciw.dists.Exponential(0.1)] does not seem to relate to an arrival rate of 6 times per hour. This distribution function requires a decimal value, so we divide the arrival rate of 6 (arrivals per hour) and divide it by 60 (the number of minutes in an hour).

Similarly, for the service time, the rate of processing per hour is 4 and is represented as $4/60 = 0.067$.

The next piece of code to add is:

Python Code

```
ciw.seed(1)
Q = ciw.Simulation(N)
# run the simulation for one shift (8 hours = 480 minutes)
Q.simulate_until_max_time(480)
```

This is an instruction to tell the computer to create a simulation and to run it for a simulated time of one shift (8 h/480 min).

That is all that is required to create the simulation. However, there are no instructions to tell the computer to report the results. The following program code does this.

Python Code

```
waitingtimes = [r.waiting_time for r in recs]
servicetimes = [r.service_time for r in recs]
avg_waiting_time = sum(waitingtimes) / len(waitingtimes)
print('Avg. wait time: ',avg_waiting_time)
avg_service_time = sum(servicetimes) / len (servicetimes)
print('Avg. processing time: ',avg_service_time)
print('Avg. machine utilisation %:',
        Q.transitive_nodes[0].server_utilisation)
```

There are three results that are reported (look for the 'print' keyword).

First, the average waiting time in minutes for each job. Second, the average time taken to process each job in minutes. Finally, the average utilisation of the machine (server) as a percentage.

When you execute your simulation you should see the following results in the console:

```
Avg. wait time:   51.51392337104136
Avg. processing time:   12.643780078229085
Avg. machine utilisation:   0.9969939361643851
```

This tells us that on average, a job took nearly 13 min to process and had to wait approximately 52 min in the queue. The machine was operating for most of the time (99.7%).

This is excellent for a shop floor supervisor who has to report the percentage of time that a machine spends idle. Hardly any downtime for the machine in this scenario.

However, let us use the simulation to start investigating different scenarios.

We shall now explore the effect of increasing the number of machines from one to two.

Edit the following line to increase the number of servers (machines) to 2:

```
number_of_servers=[2]
```

If we execute the simulation again we observe the following results:

```
Avg. wait time:   8.79660702065997
Avg. processing time:   14.249724856289776
Avg. machine utilisation:   0.6827993160305518
```

We can see that the addition of an extra machine has dramatically reduced the wait time from 52 min to around 9 min. The utilisation of the two resources has also fallen to 68%, meaning that machining resources are idle for approximately 32% of the shift.

While there is a reduction in waiting time, and therefore the overall lead time to delivery of a product, there is the additional capital cost of extra plant. Depending on how the machine is operated, there may also be extra labour required to run both machines at the same time.

The shop floor supervisor has a conversation with the company owner, and it is clear that there is no cash with which to purchase another machine. The next course of action is to try and increase the output of the timber conversion process. The service time is 15 minutes, which means that 4 jobs per hour are processed. What difference would it make if we could process 5 jobs per hour?

Edit the following line to reflect a service rate of 5 jobs per hour (5/60=0.08):

```
service_distributions=[ciw.dists.Exponential(0.08)]
```

Here are the results:

```
Avg. wait time:    26.20588722740488
Avg. processing time:   11.300597865271746
Avg. machine utilisation:   0.9485495709367999
```

The machine utilisation has increased, but the waiting time is much less than it was with a service time of 15 minutes. This illustrates that there is a significant benefit to be had by making even small changes to the service time of a process. This thinking is central to 'lean manufacturing' techniques, where potential opportunities for the removal of waste are identified. There might be some different tooling that enables the timber to be cut at a faster rate, or there might be a better way of organising the material so that the cutting-to-length operation is optimised for the fewest cuts.

5.6 Confidence

Once we have built a simulation, it is important that we are confident that it represents the situation that we are modelling.

If we look at the results we have observed so far, what do we notice about the average processing time?

We have obtained three different values: 12.6, 14.2 and 11.3 min. This is a significant variation, and it suggests that the simulation might not be taking a sufficient number of scenarios into account.

For a given scenario, there is a time when the simulation queue is empty, and then partially complete, until a steady state of operation is achieved. Similarly, towards the end of a simulations there will be a number of jobs that have not been completed. When we report the statistics of how the process has performed, we are collecting the data for jobs that have been completed.

Depending on the time require to 'wind-up' and 'wind-down' a simulation, there could be a disproportionate effect on the performance that we observe. This would decrease our confidence in ability of the simulation to be used as a tool for experimentation.

We deal with this in two ways. First, we run the simulation for a longer simulation time and then report only the performance from the system once it is in a steady state of operation. For our 8 h shift, we could add an hour before the start and at the end for warm-up and cool-down periods.

Second, we can run the simulation many times, altering a number (called a 'seed') so that each run has some variation introduced into it.

Create a new file called 'timber_conversion_2.py' and enter the following code:

Python Code

```python
import ciw

N = ciw.create_network(
    # jobs arrive every 10 minutes, or 6 times per hour
    arrival_distributions=[ciw.dists.Exponential(0.1)],
    # jobs take 15 minutes to process which is
    #    4 jobs completed per hour
    service_distributions=[ciw.dists.Exponential(0.067)],
    # the number of machines available to do the processing
    number_of_servers=[1]
)

runs = 1000 # this is the number of simulation runs
average_waits = []
average_services = []

for trial in range(runs):
    ciw.seed(trial) # change the seed for each run
    Q = ciw.Simulation(N)
    # run the simulation for one shift (8 hours = 480 minutes)
    #    + 2 hours (120 minutes)
    Q.simulate_until_max_time(600, progress_bar=True)
    recs = Q.get_all_records()
    waits = [r.waiting_time for r in recs if r.arrival_date >
        60 and r.arrival_date < 540]
    mean_wait = sum(waits) / len(waits)
    average_waits.append(mean_wait)
    services = [r.service_time for r in recs if r.arrival_date >
        60 and r.arrival_date < 540]
```

```
mean_services = sum(services) / len(services)
average_services.append(mean_services)

print('Number of simulation runs: ',runs)
print('Avg. wait time:', sum(average_waits)/len(average_waits))
print('Avg. processing time: ',
    sum(average_services)/len(average_services))
```

Execute the code and you will observe the following results:

```
Number of simulation runs:   1000
Avg. wait time: 115.69878479543915
Avg. processing time:  14.87316389724181
Avg. machine utilisation:  0.8560348867271905
```

You can now edit the variable <runs=1000> to change the number of times that the simulation executes.

At lower values for <runs> you will notice variability in the statistics that are reported. As the value of <runs> increases the statistics start to stabilise. This indicates that we can have confidence that the simulation is providing results that we can trust. This is regarded as good practice for the modelling and simulation of systems.

5.7 Conclusion

We have looked at the application of queueing to the modelling of an industrial process. This approach is straightforward to use and enables the parameters of a simulation to be specified.

This specification can then be used with a simulation tool, and in this chapter we have used Ciw to quickly construct a simulation that represents our queueing model.

As the simulation runs we can collect summary statistics that can help us understand the interrelationships between parameters such as job arrival rates, processing times and the number of resources available to do the work. We can then explore different scenarios by changing these parameters, and this can help us understand what the limits of the system might be.

Exploring different situations via simulations is an inexpensive and quick way to find the limits of a system, or to identify new possibilities. For example, you might want to find ways of increasing the output of a factory temporarily to complete a particular rush order for an important customer.

You know that you can increase capacity by adding another shift or by buying new plant. But you might want to know how many additional operators you need to bring in to complete the extra work. You'll also want to see how this might affect the impact on the rest of the orders for other customers.

A simulation can help you determine the quantity of extra resource required to complete the job, while balancing the consequences of this on the rest of the system.

While you might not be able to buy, install and commission new plant quickly enough, you would be in a better position to decide whether to outsource some of the work or not.

A more strategic consider the potential impact of the sales team's forecast for the next quarter; you could use this forecast to investigate the demands that could be made on your business resources, and then be in a position to justify the acquisition of new plant or additional staffing.

5.8 Learning Activities

? Exercise

Using the program code from above, experiment with different values. You can change the parameters for the number of simulation runs for instance, but you can also change the 'shift length'; this refers to the amount of simulated time that the program executes.

Simulation code allows us to try out different values quickly, to see what the different effects might be. This is convenient when we have a specific question to answer. However, we often need to perform deeper analysis of a a simulation model, and in such cases it is useful to record the effects of our changes.

Try to adopt good practice by recording the values that you change, noting the effects of these changes in a table. This habit will help you when your models increase in complexity.

Some good questions to ask of this model could be:

1. What is the effect on machine utilisation as the arrival rate of jobs declines?
2. How would you find an optimum set of values to ensure that the system is balanced?

? Extension exercise

When you start to build simulations, you quickly gain a deeper appreciation of the dynamics of systems. An important part of simulation is being able to discover and then communicates the results of your simulation.

Using the program code above plus the details available in Ciw documentation, develop some additional information to report.

For example, it would be interesting to see what the average length of the queue is before the machine. This will then tell us what the total inventory that is being processed amounts to (Work in Progress).

The code currently reports the average (mean) of a set of values. Enhance the reporting to include additional summary statistics such as standard deviation.

From Process to System Simulation

6

6.1 Simulating Industrial Systems

In the previous chapter we looked at the concept of queueing as an industrial process. Queues are an effective way of representing dynamic systems that receive an input, process that input and then produce an output. As such, they lend themselves to the modelling of processes or services within organisations.

When we are faced with a complex problem to be solved, we often try and simplify the representation of the object or system under scrutiny. We can look at a situation in the most abstract way possible, or we can delve deep into the system and model one tiny aspect of it in great detail.

For example, we could model an entire business with just one queue; the business would take in orders, process those orders, and we could produce a set of results using a simulation like we built in the previous chapter.

This scenario might work, in that if the order mix and the demand patterns were stable and the production processes were relatively predictable, a one-queue model would suffice. It would be limited in what it could tell us though, as there would not be a sufficient number of variables to experiment with. In the real-world, modelling an entire business as a single queue is somewhat similar to the high-level analysis that an accountant could perform with a set of spreadsheets.

We tend to be motivated to model and simulate an organisation when we feel the need to understand it better. Modelling the organisation as a set of queues is a good way of achieving improved understanding. The process of modelling itself is often enlightening, as are the questions that it prompts us to ask production personnel as we build the model.

Queues therefore allow us to model a system at different levels, and this leads to a range of opportunities to comprehend what is going on in an organisation.

Rather than abstracting the business as one queue, we could look at divisions, sections or departments, depending on how the business is subdivided. Problems often provide the impetus for improvement, and that improvement usually requires more scrutiny than we can currently perform.

© Springer Nature Switzerland AG 2021

R. Hill and S. Berry, *Guide to Industrial Analytics*, Texts in Computer Science,
https://doi.org/10.1007/978-3-030-79104-9_6

Therefore, the focus of your modelling may be a particular 'problem' process; the machining centre that cannot keep up with the rest of the factory, or the sales order processing clerk who always seems to be behind.

Queues do enable a modular approach to improve the visibility we have of a business. We can take a top-down approach where we partition the business broadly into functions that we model as separate queues. Or, we can select individual processes that we want to look at in more detail.

All of the above requires us to extend the one-queue approach to modelling of the previous chapter, so that we can look at collections of queues, both individually and as a whole. The 'whole' is the system under scrutiny, whether it be the whole business, a division, department, or a manufacturing cell. We usually refer to a collection of connected queues as a queueing network.

Fortunately, much of what we have learned in the previous chapter applies to the creation of a queueing network, so as before, new concepts will be introduced as we explore an example scenario.

6.2 Example: Joinery Manufacturer

The 'timber conversion' process from the previous chapter is a model of part of a larger manufacturing process. A joinery manufacturer manufactures make-to-order hardwood window frames for building companies. Orders are received and processed through three distinct production processes as shown in Fig. 6.1.

- Workstation 1 (WS1) is the timber conversion process modelled in the previous chapter, where lengths of wood are profiled (shaped) and cut to length.
- Workstation 2 (WS2) receives the output from Workstation 1 and prepares the ends for assembly by cutting mortice and tenon joints.
- Workstation 3 (WS3) takes the timber components and assembles them into a completed window frame, thus completing the order.

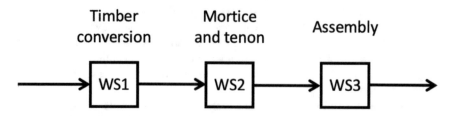

Fig. 6.1 Window frame production in a joinery factory

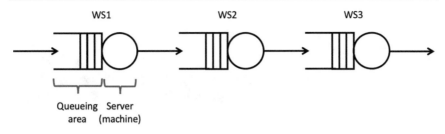

Fig. 6.2 Production line described as a series of three queues

Using a queueing approach to modelling we have three queues: one queue for each workstation. The output of WS1 feeds WS2, and the output of WS2 feeds the input of WS3. The production line is expressed as a queueing model is in Fig. 6.2.

To complete the specification of the queueing network, we need to identify the arrival and service rates, as well as the number of servers, for each queue.

? Reflection activity

What is the arrival rate for items entering WS2?

When we modelled the single queue for timber conversion in the previous example, we specified the rate at which jobs arrived, this being 6 times per hour, or every 10 mins. We also reasoned that there would be natural variability around this schedule, and therefore the actual arrival rate would be based on the exponential distribution, with a mean of 6 times per hour.

In this extended scenario, WS2 receives its jobs directly from WS1. Therefore, the arrival rate at WS2 is determined by whatever the output rate is from WS1.

The same goes for the arrival rate for WS3, who receives its input from the output from WS2.

Therefore, we should not be specifying a separate arrival rate for the downstream workstations as these are determined by how the whole queuing network operates.

However, we can specify the service (process) time for WS2 and WS3, and we can also specify the capacity of each by quantifying the number of servers available.

6.3   Building the Simulation

As before, we shall take the specification for the model and implement this as a simulation using Ciw.

First of all we create the `arrival_distributions`. We need to specify an arrival distribution for each of the queues that represent a workstation. In this example we have three workstations represented by three queues.

Table 6.1 Process times for each workstation

Workstation	Process time
WS1—Profile and cut to length	Exponential distribution with a mean of 15 mins
WS2—Mortice and tenon	Exponential distribution with a mean of 30 mins
WS3—Assemble window frame	Uniform distribution between 15 and 19 mins

The arrival rate for the first workstation is 6 times per hour. The arrival rate for the remaining queues is determined by the first queue, so we do not specify a separate arrival rate for WS2 and WS3. Instead we use the `ciw.dists.NoArrivals()` declaration.

This tells Ciw that jobs only enter the systems via WS1, and that WS2 and WS3 do not receive jobs from outside of the system.

Python Code

```
N = ciw.create_network(
    # jobs arrive at workstation 1 every 10 minutes,
    # or 6 times per hour
    # no arrivals at workstations 2 and 3
    arrival_distributions=[ciw.dists.Exponential(0.1),
                           ciw.dists.NoArrivals(),
                           ciw.dists.NoArrivals()],
```

The service distributions (range of processing times) are summarised in Table 6.1.

This is written in Ciw as follows:

Program code

```
# jobs take 15 mins to process workstation 1
# jobs take 30 mins to process workstation 2
# jobs take between 15 and 19 mins workstation
service_distributions=[ciw.dists.Exponential(0.067),
                       ciw.dists.Exponential(0.2),
                       ciw.dists.Uniform(15, 19)],
```

We now specify the servers for each of the queues. As before there is one machine at Workstation 1. There is one machine at Workstation 2. Workstation 3

Table 6.2 Operations for each workstation

Workstation	Operations
WS1—Profile and cut to length	Shape timber for cill profile Shape timber for jamb profile Shape timber for rail profile Cut cill section to length Cut jamb section to length Cut rail section to length
WS2—Mortice and tenon	Mortice cill section Mortice top rail section Tenon side rail section Tenon jamb section
WS3—Assemble window frame	Glue components Assemble frame and cramp Affix beading

is where the timber components are assembled into a completed frame. This area contains three people and thus is represented by 3 servers. As we have specified individual arrival and service distributions, we also declare the servers with the `number_of_servers` statement:

```
number_of_servers=[1, 1, 3]
```

The last bit of information to complete the model is to specify the routing of the job. A production routing is a way of describing how an item is processed through different operations. The production routing for this scenario is shown in Table 6.2.

For the simulation, we need to specify the order in which the queues are executed. From the production routing the order progresses from WS1, through WS2 to WS3.

The routing is specified as an $n x n$ matrix, where n = the number of nodes in the network. Each node represents a separate queue, which in our example equates to WS1–WS3.

In turn, we take each queue and identify the probability that a queue transitions from the previous one. So, for the first queue in our example (WS1) we compile a set of values as follows:

```
WS1: [0.0, 1.0, 0.0]
```

The probability that node 1 transitions to itself (WS1 to WS1) is zero. The probability that node 1 transitions to node 2 (WS1 to WS2) is 1.0, or 100%. The probability that node 1 transitions to node 3 (WS1 to WS3) is zero.

For WS2 we have:

```
WS2: [0.0, 0.0, 1.0]
```

The probability that node 2 transitions to node 1 (WS2 to WS1) is zero. The probability that node 2 transitions to node 2 (WS2 to WS2) is zero. The probability that node 2 transitions to node 3 (WS2 to WS3) is 100%.

For WS3 we have:

```
WS3: [0.0, 0.0, 0.0]
```

The probability that node 3 transitions to node 1 (WS3 to WS1) is zero. The probability that node 3 transitions to node 2 (WS3 to WS2) is zero. The probability that node 3 transitions to node 3 (WS3 to WS3) is zero.

Each of the lists of values for the individual workstations is now brought together into a matrix:

```
[0.0, 1.0, 0.0]
[0.0, 0.0, 1.0]
[0.0, 0.0, 0.0]
```

This example has shown an exhaustive approach to arriving at the routing matrix for the system model. In practice, we locate the points of transition for a particular routing and insert '1.0', which would give:

```
[       1.0      ]
[             1.0]
[                ]
```

The remaining spaces are then completed with zeroes. This becomes intuitive after a couple of attempts.

Finally, we should recognise that as the simulation stands, there is an assumption that the queue capacity is infinite—there is no limit on the number of jobs that can wait for an operation. This is not realistic, and organisations often have restricted capacity for the storage of items. In Ciw we use `queue_capacities=[]` to specify a limit for each of the queues in the system. For this system we shall specify a maximum queue length of 3 jobs for each process.

Program code

```
N = ciw.create_network(
    # jobs arrive at workstation 1 every 10 minutes,
    # or 6 times per hour
    # no arrivals at workstations 2 and 3
    arrival_distributions=[ciw.dists.Exponential(0.1),
                           ciw.dists.NoArrivals(),
                           ciw.dists.NoArrivals()],

    # jobs take 15 mins to process workstation 1
    # jobs take 30 mins to process workstation 2
    # jobs take between 8 and 15 mins workstation
```

```
service_distributions=[ciw.dists.Exponential(0.067),
                       ciw.dists.Exponential(0.2),
                       ciw.dists.Uniform(15, 19)],
# the number of machines available to do the processing
number_of_servers=[1, 1, 3],
routing=[[0.0, 1.0, 0.0],
         [0.0, 0.0, 1.0],
         [0.0, 0.0, 0.0]],
queue_capacities=[3, 3, 3]
)
```

Even with only two additional queues on top of the single queue in the previous chapter, the system is becoming more complex to visualise.

Communicating the performance of the model as well as the results of any experimentation are a fundamental part of simulation. We need to see the results so that we can verify the suitability of a model, but we may also need something to show what we have discovered to others. If communication of the insights is hampered, it is unlikely that any resulting course of action will garner support from the relevant stakeholders.

Since we are simulating a system of three workstations, it is likely that we shall be interested in the utilisation of each of those workstations in relation to each other, which suggests at least one type of visualisation.

We can report the overall time that a job spends queueing and being processed. It would also be useful to count the quantity of jobs that have been completed at the end of a shift.

Again, as in the previous example we should include the ability to conduct a number of experiments to explore what the steady-state performance of the system is. We still might want to investigate what a 'cold-start' looks like though, for instance if the shift starts and the system has no jobs in progress (or 'work in progress' (WIP))as it is typically referred to).

As the number of runs increases it is useful to have some feedback to the console that the simulation is working, otherwise there may be nothing to see for a while. We can obtain this feedback by adding `progress_bar=True` to the main simulation call:

```
Q.simulate_until_max_time(600, progress_bar=True)
```

To assist with visualisation of the results, the simulation produces a bar chart of the proportion of each workstation that has been utilised during the simulation, as per Fig. 6.3.

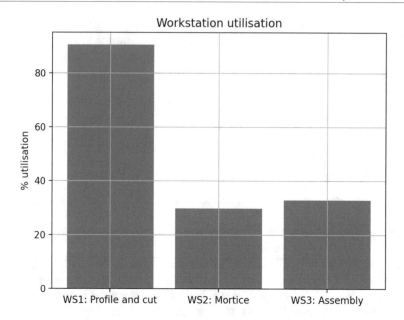

Fig. 6.3 Resource utilisation for each workstation

From the first run of the simulation there is an imbalance of activity across the workstations.

```
Number of simulation runs:   100
Avg. wait time: 10.662043705489934
Avg. processing time:  12.372948802060852
Avg. jobs completed in shift:   31
Avg. util WS1:   91 %
Avg. util WS2:   30 %
Avg. util WS3:   33 %
```

? Question

A total of 31 jobs were completed during the shift. How would you change the simulation to investigate ways of achieving a more balanced distribution of work?

The complete program listing is as follows:

Python Code

```python
import ciw
import math
import matplotlib.pyplot as plt

N = ciw.create_network(
    # jobs arrive at workstation 1 every 10 minutes,
    # or 6 times per hour
    # no arrivals at workstations 2 and 3
    arrival_distributions=[ciw.dists.Exponential(0.1),
                           ciw.dists.NoArrivals(),
                           ciw.dists.NoArrivals()],

    # jobs take 15 mins to process workstation 1
    # jobs take 30 mins to process workstation 2
    # jobs take between 8 and 15 mins workstation
    service_distributions=[ciw.dists.Exponential(0.067),
                           ciw.dists.Exponential(0.2),
                           ciw.dists.Uniform(15, 19)],
    # the number of machines/resources
    # available to do the processing
    number_of_servers=[1, 1, 3],
    routing=[[0.0, 1.0, 0.0],
             [0.0, 0.0, 1.0],
             [0.0, 0.0, 0.0]],
    queue_capacities=[3, 3, 3]
)

runs = 100 # this is the number of simulation runs

# declaring some empty lists to record values into
average_waits = []
average_services = []
completed_jobs=[]
ws1_util = []
ws2_util = []
ws3_util = []
node_util = []

# main simulation loop governed
# by (runs)
for trial in range(runs):
```

```
ciw.seed(trial) # change the seed for each run
Q = ciw.Simulation(N)

# run the simulation for one shift
# (8 hours = 480 minutes) + 2 hours (120 minutes)
# progress bar gives feedback for longer runs
Q.simulate_until_max_time(600, progress_bar=True)
recs = Q.get_all_records()

# obtain the utilisation for each
# of the nodes/workstations
ws1_util.append(Q.transitive_nodes[0].server_utilisation)
ws2_util.append(Q.transitive_nodes[1].server_utilisation)
ws3_util.append(Q.transitive_nodes[2].server_utilisation)

# calculate some summary statistics
waits = [r.waiting_time for r in recs if r.arrival_date
> 60 and r.arrival_date < 540]
mean_wait = sum(waits) / len(waits)
average_waits.append(mean_wait)
services = [r.service_time for r in recs if r.arrival_date
> 60 and r.arrival_date < 540]
mean_services = sum(services) / len(services)
average_services.append(mean_services)
num_completed = len([r for r in recs if r.node
==3 and r.arrival_date < 540])
completed_jobs.append(num_completed)

# report results to the console
print('Number of simulation runs: ',runs)
print('Avg. wait time:', sum(average_waits)/
    len(average_waits))
print('Avg. processing time: ',
    sum(average_services)/len(average_services))
print('Avg. jobs completed in shift: ',
    math.floor(sum(completed_jobs) /
    len(completed_jobs)))
avgWS1util = sum(ws1_util) / len(ws1_util)
print('Avg. util WS1: ', round(avgWS1util*100), '%')
node_util.append(avgWS1util*100)
avgWS2util = sum(ws2_util) / len(ws2_util)
print('Avg. util WS2: ',round(avgWS2util*100), '%')
node_util.append(avgWS2util*100)
avgWS3util = sum(ws3_util) / len(ws3_util)
print('Avg. util WS3: ',round(avgWS3util*100), '%')
```

```
node_util.append(avgWS3util*100)

# produce histogram of resource utilisation
plt.bar(['WS1: Profile and cut', 'WS2: Mortice',
    'WS3: Assembly'],
    node_util)
plt.grid(True)
plt.ylabel('% utilisation')
plt.title('Workstation utilisation')
plt.show()
```

6.4 Managing Resource Utilisation

In this example of a manufacturing system we have seen that there is an inherent imbalance of resource utilisation. Even if the capacity of the assembly workstation is reduced to 1 person (saving two staff), the utilisation of workstation 2 still only amounts to approximately 30%.

This presents at least three opportunities for the business to consider. First, can the operator of workstation 2 be redeployed for some of the 70% of the time that they appear to be idle?

Second, can the machinery of WS1 be modified to include jointing as part of the process. This would extend the processing time at WS1, but would also eliminate the queue time and process time at WS2.

Third, can the business find new orders that require processing at WS2, but not WS1? For example, can pre-shaped lengths of timber be bought in, bypassing WS1?

Opportunity one would probably warrant some investigation with the production staff. There may be some training and reskilling required to explore before any further experiments can be conducted.

Similarly, opportunity two would require some research into the feasibility of combining the first two workstations. There is likely to be a cost associated with this approach and that is where the simulation can help identify whether sufficient value can be achieved.

The third opportunity is an interesting scenario. There may be a market that exists whereby other timber conversion operations require mortice and tenon jointing as a service. Or, there may be a supplier that can deliver ready-shaped blanks that are in the same state as those that are produced by WS1. The business could therefore consider buying ready-shaped blanks in standard sizes and dedicating the in-house WS1 to be reserved for bespoke jobs, with custom sizes.

To explore the potential of the situation, we need to look at how we might modify the model of the system.

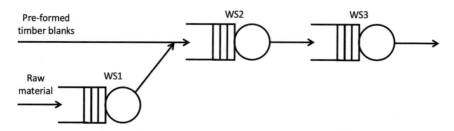

Fig. 6.4 Modified queueing model to reflect the introduction of pre-formed timber blanks, which are processed along with timber blanks produced by WS1, at WS2

Figure 6.4 shows how the model needs to change. WS1 accepts raw timber and the output from this goes straight to WS2 as before. WS2 now accepts pre-shaped timber blanks directly from an external supplier in addition to the output from WS1.

As far as the product routing is concerned there are no changes required. The only change is to add an arrival rate to WS2 so that it can accept the blanks.

We therefore modify the simulation as follows:

Python Code

```
N = ciw.create_network(
    # jobs arrive at workstation 1 every 10 minutes,
    # or 6 times per hour
    # shaped blanks arrive at WS 2
    # at 6 times per hour
    # no arrivals at workstation 3
    arrival_distributions=[ciw.dists.Exponential(0.1),
                           ciw.dists.Exponential(0.1),
                           ciw.dists.NoArrivals()],
```

The difference in resource utilisation is summarised in Fig. 6.5 and Table 6.3.

```
Number of simulation runs:  100
Avg. wait time: 10.358355470577324
Avg. processing time:  11.676648501518232
Avg. jobs completed in shift:  75
Avg. util WS1:  91 %
Avg. util WS2:  72 %
Avg. util WS3:  77 %
```

The addition of bought-in pre-shaped timber blanks to WS2 enables the spare capacity to be utilised to a greater extent. This has the additional benefit of increasing the

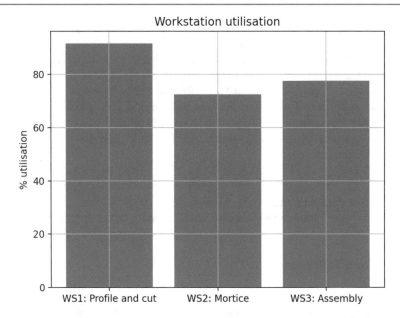

Fig. 6.5   Resource utilisation for each workstation after introducing pre-shaped timber blanks

Table 6.3   Workstation utilisation before and after introduction of bought-in pre-formed timber blanks

Resource utilisation	Resource utilisation with bought-in blanks
WS1: 91	WS1: 91
WS2: 30	WS2: 72
WS3: 32	WS3: 77

flow of materials into WS3, resulting in the number of jobs completed each shift rising from 31 to 75. Overall, the resource utilisation is better balanced across the system.

6.5   Product Mixes

A common query for a manufacturing plant is whether the sales function should sell a particular type of product, or perhaps run a sales promotion. The product that is sold might make the least demand on the factory, or it might balance the utilisation of resources when combined with orders of other types.

Alternatively there might be a desire to examine the feasibility of adding a new product type to the existing portfolio, to assess what the potential for profitability might be. This can be helpful for businesses that experience variability in sales in relation to the season.

For instance, a gas heating boiler manufacturer might expect to sell more units during the winter period and might therefore make to stock during the summer period. If the boiler manufacturer added a gas barbecue to its product line, it might expect to sell these during spring and summer and therefore could make these items for stock during the winter months.

This strategy would keep the factory work level more stable by manufacturing different (but similar) products that require the same manufacturing resources to produce. We shall now extend the window manufacturing example to investigate what the impact of introducing a new product might be on the factory resources.

6.5.1 Sash Windows

The sales team have identified a potential new market for sash windows, which are different to the window frames that are made at present. Most of the manufacturing processes are similar between the two products, with the addition of a workstation to machine the sashes. These products will not be processed by the existing morticing workstation, but they will take longer to assemble at the last stage of manufacture.

The proposed factory layout now requires four workstations in total as shown in Fig. 6.6. Existing windows follow the established route. Sash windows require a different process at stage two of their manufacture and are routed to WS3 before returning to be assembled alongside the plain window frames at WS4. The proposed routings are given in Table 6.4. As before, we shall consider the arrival distributions for our simulation.

Plain window frame orders arrive at WS1 every 10 min. Sash windows arrive at WS1 every 30 min.

In the previous examples we have specified arrival distributions for each of the workstations in the model, but we have assumed only one type of product. We are now

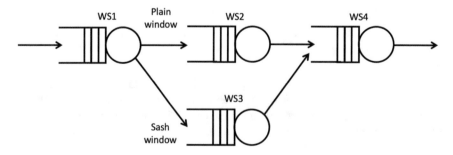

Fig. 6.6 Production line modified to manufacture two distinct types of product

Table 6.4 Production routings for plain window frames and sash window frames

Workstation	Plain window frame	Sash window frame
WS1—Profile and cut	X	X
WS2—Mortice	X	
WS3—Sash		X
WS4—Assemble	X	X

interested in two discrete product types. Ciw has the concept of 'customer classes' which we use to identify different product types in this simulation. All classes are numbered from zero upwards.

We shall assume that plain window frames are 'Class 0' and sash windows are represented by 'Class 1'.

We also assume that we are only modelling products that are made totally in-house, so the remaining workstations (WS2, WS3 and WS4) have 'NoArrivals()' as their arrival distributions.

Python Code

```
N = ciw.create_network(
    # Plain window jobs arrive at
    # WS1 every 10 minutes
    # or 6 times per hour
    # Sash window jobs arrive at
    # WS1 every 30 minutes
    # at twice per hour
    # no arrivals at WS2 (Mortice),
    # WS3 (Sash), or WS4 (Assembly)
    arrival_distributions={'Class 0': [ciw.dists.Exponential(0.1),
                                       ciw.dists.NoArrivals(),
                                       ciw.dists.NoArrivals(),
                                       ciw.dists.NoArrivals()],
                           'Class 1': [ciw.dists.Exponential(0.1),
                                       ciw.dists.NoArrivals(),
                                       ciw.dists.NoArrivals(),
                                       ciw.dists.NoArrivals()]},
```

The arrival distributions are declared independently for each customer class. We shall now specify the service distributions.

For the processing time, we know that plain window frames are not processed by WS3, so we declare a dummy distribution using `ciw.dists.Deterministic(0.0)` for node 3.

Similarly we do the same for node 2 (WS2) for the sash window, that does not pass through this process.

Python Code

```
# Plain windows take 15 mins WS1
# Sash windows take 30 mins WS1
# Plain windows take 5 mins WS2
# Sash windows take 15 mins WS3
# Plain windows: between 8 and 15 mins WS4
# Sash windows: between 12 and 20 mins WS4
service_distributions={'Class 0': [ciw.dists.Exponential(0.067),
                                   ciw.dists.Exponential(0.2),
                                   ciw.dists.Deterministic(0.0),
                                   ciw.dists.Uniform(15, 19)],
                        'Class 1': [ciw.dists.Exponential(0.03),
                                   ciw.dists.Deterministic(0.0),
                                   ciw.dists.Exponential(0.067),
                                   ciw.dists.Uniform(22, 40)]},
```

We then specify the remainder of the simulation set-up.

We assume a capacity of 1 for each WS to start with. Each of the products is routed differently, and this is declared in the 'routing' section, as before for each of the customer classes.

Finally some queue capacities are set to constrain how much product material is drawn into the factory as WIP.

Python Code

```
# the number of machines/resources
# available to do the processing
number_of_servers=[1, 1, 1, 1],
routing={'Class 0':[[0.0, 1.0, 0.0, 0.0],
                    [0.0, 0.0, 0.0, 1.0],
                    [0.0, 0.0, 0.0, 0.0],
                    [0.0, 0.0, 0.0, 0.0]],
          'Class 1':[[0.0, 0.0, 1.0, 0.0],
                    [0.0, 0.0, 0.0, 0.0],
                    [0.0, 0.0, 0.0, 1.0],
                    [0.0, 0.0, 0.0, 0.0]]},
queue_capacities=[3, 3, 3, 3]
```

The remainder of the simulation is mostly as before, with the addition of reporting for the additional workstation.

Python Code

```
import ciw
import math
import matplotlib.pyplot as plt

N = ciw.create_network(
    # Plain window jobs arrive at
    # WS1 every 10 minutes
    # or 6 times per hour
    # Sash window jobs arrive at
    # WS1 every 30 minutes
    # at twice per hour
    # no arrivals at WS2 (Mortice),
    # WS3 (Sash), or WS4 (Assembly)
    arrival_distributions={'Class 0': [ciw.dists.Exponential(0.1),
                                       ciw.dists.NoArrivals(),
                                       ciw.dists.NoArrivals(),
                                       ciw.dists.NoArrivals()],
                           'Class 1': [ciw.dists.Exponential(0.1),
                                       ciw.dists.NoArrivals(),
                                       ciw.dists.NoArrivals(),
                                       ciw.dists.NoArrivals()]},
    # Plain windows take 15 mins WS1
    # Sash windows take 30 mins WS1
    # Plain windows take 5 mins WS2
    # Sash windows take 15 mins WS3
    # Plain windows: between 8 and 15 mins WS4
    # Sash windows: between 12 and 20 mins WS4
    service_distributions={'Class 0': [ciw.dists.Exponential(0.067),
                                       ciw.dists.Exponential(0.2),
                                       ciw.dists.Deterministic(0.0),
                                       ciw.dists.Uniform(15, 19)],
                           'Class 1': [ciw.dists.Exponential(0.03),
                                       ciw.dists.Deterministic(0.0),
                                       ciw.dists.Exponential(0.067),
                                       ciw.dists.Uniform(22, 40)]},
    # the number of machines/resources
    # available to do the processing
    number_of_servers=[1, 1, 1, 1],
    routing={'Class 0':[[0.0, 1.0, 0.0, 0.0],
                        [0.0, 0.0, 0.0, 1.0],
                        [0.0, 0.0, 0.0, 0.0],
                        [0.0, 0.0, 0.0, 0.0]],
             'Class 1':[[0.0, 0.0, 1.0, 0.0],
                        [0.0, 0.0, 0.0, 0.0],
                        [0.0, 0.0, 0.0, 1.0],
                        [0.0, 0.0, 0.0, 0.0]]},
    queue_capacities=[3, 3, 3, 3]
)

runs = 100 # this is the number of simulation runs

# declaring some empty lists to record values into
average_waits = []
average_services = []
```

```
completed_jobs=[]
ws1_util = []
ws2_util = []
ws3_util = []
ws4_util = []
node_util = []

# main simulation loop governed
# by (runs)
for trial in range(runs):
    ciw.seed(trial) # change the seed for each run
    Q = ciw.Simulation(N)

    # run the simulation for one shift
    # (8 hours = 480 minutes) + 2 hours (120 minutes)
    # progress bar gives feedback for longer runs
    Q.simulate_until_max_time(600, progress_bar=True)
    recs = Q.get_all_records()

    # obtain the utilisation for each
    # of the nodes/workstations
    ws1_util.append(Q.transitive_nodes[0].server_utilisation)
    ws2_util.append(Q.transitive_nodes[1].server_utilisation)
    ws3_util.append(Q.transitive_nodes[2].server_utilisation)
    ws4_util.append(Q.transitive_nodes[3].server_utilisation)
    # calculate some summary statistics
    waits = [r.waiting_time for r in recs if r.arrival_date
        > 60 and r.arrival_date < 540]
        mean_wait = sum(waits) / len(waits)
    average_waits.append(mean_wait)
    services = [r.service_time for r in recs if
        r.arrival_date > 60 and r.arrival_date < 540]
        mean_services = sum(services) / len(services)
    average_services.append(mean_services)
    num_completed = len([r for r in recs if r.node ==3
        and r.arrival_date < 540])
    completed_jobs.append(num_completed)

# report results to the console
print('Number of simulation runs: ',runs)
print('Avg. wait time:', sum(average_waits)/
    len(average_waits))
print('Avg. processing time: ', sum(average_services)
    /len(average_services))
print('Avg. jobs completed in shift: ',
    math.floor(sum(completed_jobs) / len(completed_jobs)))
avgWS1util = sum(ws1_util) / len(ws1_util)
print('Avg. util WS1: ', round(avgWS1util*100), '%')
node_util.append(avgWS1util*100)
avgWS2util = sum(ws2_util) / len(ws2_util)
print('Avg. util WS2: ',round(avgWS2util*100), '%')
node_util.append(avgWS2util*100)
avgWS3util = sum(ws3_util) / len(ws3_util)
print('Avg. util WS3: ',round(avgWS3util*100), '%')
node_util.append(avgWS3util*100)
avgWS4util = sum(ws4_util) / len(ws4_util)
print('Avg. util WS4: ',round(avgWS4util*100), '%')
```

```
node_util.append(avgWS4util*100)

# produce histogram of resource utilisation
plt.bar(['WS1: Prof. and cut', 'WS2: Mortice', 'WS3: Sash',
    'WS4: Assembly'], node_util)
plt.grid(True)
plt.ylabel('% utilisation')
plt.title('Workstation utilisation')
plt.show()
```

After running the simulation you should see the following results:

```
Number of simulation runs:  100
Avg. wait time: 38.11181105045973
Avg. processing time:  19.365007197148078
Avg. jobs completed in shift:  10
Avg. util WS1:  99 %
Avg. util WS2:  25 %
Avg. util WS3:  43 %
Avg. util WS4:  80 %
```

These numbers and the associated bar chart illustrate a resource utilisation imbalance similar to what we observed in the previous example. At that stage we added machined timber blanks to bypass WS1 and go straight for morticing. As before, we can experiment with different input parameters to see how the system behaves.

One of the associated questions that is common from sales functions is 'what is the lead time to delivery of product X?'

Our simulation does not report this at present and as any production supervisor knows that if we manage a factory to maximise the resource utilisation, we shall find it more difficult to produce items on time for customers.

Since we have specified our products as separate classes in Ciw, we can report the average time that the product spends waiting in the system.

We append the following to the main simulation loop:

```
# insert after summary statistics code
# calculate waiting times for each product type
waits_1 = [r.waiting_time for r in recs if
        r.customer_class == 0]
waits_2 = [r.waiting_time for r in recs if
        r.customer_class == 1]
average_cust_waits_1.append(sum(waits_1) /
        len(waits_1))
average_cust_waits_2.append(sum(waits_2) /
        len(waits_2))
```

This calculates the average time that each product spends waiting within the system.

We can report this to the console as follows:

```
# calculate mean of all runs for each class
mean1 = sum(average_cust_waits_1) / len(average_cust_waits_1)
mean2 = sum(average_cust_waits_2) / len(average_cust_waits_2)
print('Plain window frame wait time: ', round(mean1),'mins')
print('Sash window frame wait time: ', round(mean2),'mins')
```

Finally, it has just occurred to you that there is no confirmation that you have specified the routing correctly.

How can you be sure that the individual products go to the relevant workstations for processing?

In the following code we test each class to see if it is present at the nodes which represent the correct route. For the plain window frames this means that material arrives at WS1, WS2 and WS4, which are represented by nodes 1, 2 and 4, respectively.

Sash windows are routed through WS1, WS3 and WS4, giving nodes 1, 3 and 4. If the products are routed correctly, 'True' is returned to the console.

You can experiment with the lists `visited_by_plain_window` and `visited_by_sash_window` to verify the routings.

```
# check that products are routed correctly
visited_by_plain_window = {1, 2, 4}
print('Plain windows routed through WS2? ',
        set([r.node for r in recs if r.customer_class == 0])
        == visited_by_plain_window)
visited_by_sash_window = {1, 3, 4}
print('Sash windows routed through WS3? ',
        set([r.node for r in recs if r.customer_class == 1])
        == visited_by_sash_window)
```

The final program code listing is:

Python Code

```
import ciw
import math
import matplotlib.pyplot as plt

N = ciw.create_network(
    # Plain window jobs arrive at
    # WS1 every 10 minutes
    # or 6 times per hour
    # Sash window jobs arrive at
```

```
        # WS1 every 30 minutes
        # at twice per hour
        # no arrivals at WS2 (Mortice),
        # WS3 (Sash), or WS4 (Assembly)
        arrival_distributions={'Class 0': [ciw.dists.Exponential(0.1),
                                           ciw.dists.NoArrivals(),
                                           ciw.dists.NoArrivals(),
                                           ciw.dists.NoArrivals()],
                               'Class 1': [ciw.dists.Exponential(0.1),
                                           ciw.dists.NoArrivals(),
                                           ciw.dists.NoArrivals(),
                                           ciw.dists.NoArrivals()]},
        # Plain windows take 15 mins WS1
        # Sash windows take 30 mins WS1
        # Plain windows take 5 mins WS2
        # Sash windows take 15 mins WS3
        # Plain windows: between 8 and 15 mins WS4
        # Sash windows: between 12 and 20 mins WS4
        service_distributions={'Class 0': [ciw.dists.Exponential(0.067),
                                           ciw.dists.Exponential(0.2),
                                           ciw.dists.Deterministic(0.0),
                                           ciw.dists.Uniform(15, 19)],
                               'Class 1': [ciw.dists.Exponential(0.03),
                                           ciw.dists.Deterministic(0.0),
                                           ciw.dists.Exponential(0.067),
                                           ciw.dists.Uniform(22, 40)]},
        # the number of machines/resources
        # available to do the processing
        number_of_servers=[1, 1, 1, 1],
        routing={'Class 0': [[0.0, 1.0, 0.0, 0.0],
                             [0.0, 0.0, 0.0, 1.0],
                             [0.0, 0.0, 0.0, 0.0],
                             [0.0, 0.0, 0.0, 0.0]],
                 'Class 1': [[0.0, 0.0, 1.0, 0.0],
                             [0.0, 0.0, 0.0, 0.0],
                             [0.0, 0.0, 0.0, 1.0],
                             [0.0, 0.0, 0.0, 0.0]]},
        queue_capacities=[3, 3, 3, 3]
)

runs = 100 # this is the number of simulation runs

# declaring some empty lists to record values into
average_waits = []
average_services = []
completed_jobs=[]
ws1_util = []
ws2_util = []
ws3_util = []
ws4_util = []
node_util = []
average_cust_waits_1 = []
average_cust_waits_2 = []

# main simulation loop governed
# by (runs)
for trial in range(runs):
```

```
ciw.seed(trial) # change the seed for each run
Q = ciw.Simulation(N)

# run the simulation for one shift
# (8 hours = 480 minutes) + 2 hours (120 minutes)
# progress bar gives feedback for longer runs
Q.simulate_until_max_time(600, progress_bar=True)
recs = Q.get_all_records()

# obtain the utilisation for each
# of the nodes/workstations
ws1_util.append(Q.transitive_nodes[0].server_utilisation)
ws2_util.append(Q.transitive_nodes[1].server_utilisation)
ws3_util.append(Q.transitive_nodes[2].server_utilisation)
ws4_util.append(Q.transitive_nodes[3].server_utilisation)

# calculate some summary statistics
waits = [r.waiting_time for r in recs if r.arrival_date
        > 60 and r.arrival_date < 540]
mean_wait = sum(waits) / len(waits)
average_waits.append(mean_wait)
services = [r.service_time for r in recs if r.arrival_date
          > 60 and r.arrival_date < 540]
mean_services = sum(services) / len(services)
average_services.append(mean_services)
num_completed = len([r for r in recs if r.node ==3
        and r.arrival_date < 540])
completed_jobs.append(num_completed)

# calculate waiting times for each product type
waits_1 = [r.waiting_time for r in recs
        if r.customer_class == 0]
waits_2 = [r.waiting_time for r in recs
        if r.customer_class == 1]
average_cust_waits_1.append(sum(waits_1) /
        len(waits_1))
average_cust_waits_2.append(sum(waits_2) /
        len(waits_2))

# report results to the console
print('Number of simulation runs: ',runs)
print('Avg. wait time:',
    round(sum(average_waits)/len(average_waits)),'mins')
print('Avg. processing time: ',
    round(sum(average_services)/len(average_services)),'mins')
print('Avg. jobs completed in shift: ',
    math.floor(sum(completed_jobs) / len(completed_jobs)))
avgWS1util = sum(ws1_util) / len(ws1_util)
print('Avg. util WS1: ', round(avgWS1util*100), '%')
node_util.append(avgWS1util*100)
avgWS2util = sum(ws2_util) / len(ws2_util)
print('Avg. util WS2: ',round(avgWS2util*100), '%')
node_util.append(avgWS2util*100)
avgWS3util = sum(ws3_util) / len(ws3_util)
print('Avg. util WS3: ',round(avgWS3util*100), '%')
node_util.append(avgWS3util*100)
avgWS4util = sum(ws4_util) / len(ws4_util)
```

```
print('Avg. util WS4: ',round(avgWS4util*100), '%')
node_util.append(avgWS4util*100)

# calculate mean of all runs for each class
mean1 = sum(average_cust_waits_1) / len(average_cust_waits_1)
mean2 = sum(average_cust_waits_2) / len(average_cust_waits_2)
print('Plain window frame wait time: ', round(mean1),'mins')
print('Sash window frame wait time: ', round(mean2),'mins')

# check that products are routed correctly
visited_by_plain_window = {1, 2, 4}
print('Plain windows routed through WS2? ',
        set([r.node for r in recs if r.customer_class == 0])
        == visited_by_plain_window)
visited_by_sash_window = {1, 3, 4}
print('Sash windows routed through WS3? ',
        set([r.node for r in recs if r.customer_class == 1])
        == visited_by_sash_window)

# produce histogram of resource utilisation
plt.bar(['WS1: Prof. and cut', 'WS2: Mortice',
        'WS3: Sash', 'WS4: Assembly'], node_util)
plt.grid(True)
plt.ylabel('% utilisation')
plt.title('Workstation utilisation')
plt.show()
```

Running the simulation will give the following output:

```
Number of simulation runs:  100
Avg. wait time: 38 mins
Avg. processing time:  19 mins
Avg. jobs completed in shift:  10
Avg. util WS1:   99 %
Avg. util WS2:   25 %
Avg. util WS3:   43 %
Avg. util WS4:   80 %
Plain window frame wait time:  36 mins
Sash window frame wait time:  35 mins
Plain windows routed through WS2?  True
Sash windows routed through WS3?  True
```

6.6 Conclusion

This chapter builds on previous work and illustrates the applicability of queueing networks for modelling systems of linked multiple processes.

Organisational systems can be modelled at various levels of abstraction, depending on the detail of scrutiny that is required. Very complex systems may benefit from high-level abstract models to help understand where more detail investigation is required.

As the complexity of a model increases, the visualisation of system performance becomes more important, especially so that any insight can be represented accurately and prominently. The audience should not be expected to wade through excessive information when that could have been presented graphically. Communication of the findings of simulation experiments is vital if there is a need to persuade a stakeholder to invest in a proposed solution.

Finally, once a model has been created, it is simple to modify the model to accommodate experiments that would otherwise not be practical. We can enquire as to the impact of the purchase of new plant or even new ways of doing business in the relative safety of a computer simulation. The results of the simulation should help rule out costly mistakes.

6.7 Learning Activities

Using the last model, experiment with different combinations of queue capacity, as well as the work capacity of WS3 (number of servers) to optimise the balance of work across the workstations.

? Exercise

Using the last simulation model, add the pre-machined blanks to the simulation (*hint:* think about arrival rates at the relevant workstation).

Experiment with server and queue capacities to see how this affects the waiting time within the system for each product type.

? Extension exercise

Extend the model to include a timber treatment workstation, that is used to remove any moisture from the assembled timber window frame and then inject a preservative. This would become WS5.

Not all of the orders require this treatment. You have two options for the modelling of this.

First, you can assume that a certain proportion of the assembled items require treatment, and therefore you could specify a probabilistic routing to WS5 from WS4.

Second, you could specify additional product types (customer classes) for the products that require the treatment.

What are the advantages and potential disadvantages of either of these approaches?

Think about the ease of modelling versus the accuracy (or realism) of the simulation results.

Which model might be more convincing?

Constructing Machine Learning Models for Prediction

7

7.1 Introduction

With the vast increase and availability of data, machine learning (ML) has become extremely popular in the twenty-first century and it is extensively used in many diverse fields such as computing, biology, physics, politics and agriculture. Broadly speaking machine learning is a set of techniques which allow a machine to be trained to make reliable predictions/decisions from exposure to collated data, e.g. previous daily sales of a product.

While statistical methods such as forecasting have a long and documented history of use in demand planning [1] due to the ease of data collection, they have less frequently been used in the production planning stage. Traditional methods such as linear programming have been employed with much success in the production planning and supply chain management [2–4]; however, the rise of cheap sensors to automate the collection of data has opened up the opportunity for the use of predictive methods to be used in conjunction with factory simulations [5]. As an example, better job time prediction can lead to better customer satisfaction and, should a firm be part of a supply chain, lead to better estimates that can be used for supply chain optimisation which can lead to benefits to some or all partners associated with the supply chain.

In this chapter we explore the use of such a workflow to predict job times through sensor capture, simulation and supervised machine learning regression. It is also the aim to clearly explain the process in a concise manner and showcase some simple machine learning techniques and their potential applicability. To this end a very simple hypothetical small factory will be employed as an example throughout the chapter. All data generated for the examples was done using SimPy. The datasets can be downloaded from https://bit.ly/3DUmD12, and worked examples are given at the end of the chapter.

© Springer Nature Switzerland AG 2021
R. Hill and S. Berry, *Guide to Industrial Analytics*, Texts in Computer Science,
https://doi.org/10.1007/978-3-030-79104-9_7

7.2 Data and Prediction

Analysing data and making predictions are something that all people are familiar with. On a conscious or subconscious level we make predictions based on past experience or knowledge of data everyday. For example, what time should I leave to arrive on time for work? Is that a reasonable price to pay for my lunchtime sandwich? Both of these involve past experience (data) and an ability to extrapolate trends from experiences to make a prediction, which subsequently guides our decision-making. This can be summarised by the diagram in Fig. 7.1, one should note that the internal model that we use is constantly being updated by new data and Fig. 7.1 actually showcases the point in time at which the prediction and decision are made. Supervised regression models essentially act in this way, we assume that a relationship exists between features of the data, data is then used to train a model which can then make predictions about future data points based on a set of features. The simplest example of this is a linear relationship between two features, e.g. height and weight (Fig. 7.2). Based on the straight line model we can make predictions about a person's weight given their current height. In machine learning we refer to a set of features as $x = (x_1, x_2, \ldots, x_n)$ and the observed outputs y as the targets. The underlying relationship is described by $y = f(x) + \epsilon$, where f is the real unknown model and ϵ is some noise from unknown factors and natural stochastic noise. The job of supervised learning is to build a predictive model $\hat{f}$ that closely resembles f, new predictions are labelled $\hat{y}$, this is shown in Fig. 7.3. Therefore, the job of a model can be seen as producing an output $\hat{y}$ that as closely resembles y as possible. In the height and weight example, height would be the features (x), weight the target (y) and the straight line predictor would be our predictive model $\hat{f}$ (Fig. 7.2).

A simple description of the steps needed to build a machine learning model are:

1. Collect the relevant data
2. Split the data into a training set and test set
3. Train a model on the training set
4. Use the model to make predictions on the test set
5. Compare these predictions with the actual test set for accuracy and validation purposes.

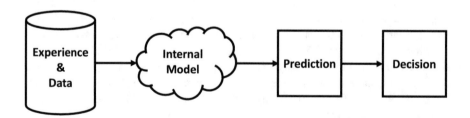

Fig. 7.1 Simple depiction of human prediction

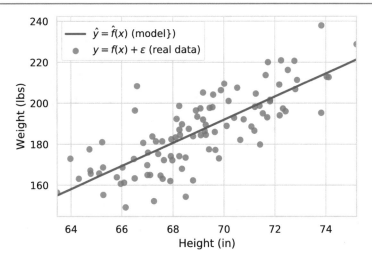

Fig. 7.2 Height 'versus' weight plotted with linear regression model

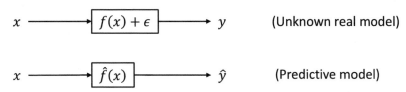

Fig. 7.3 Unknown real model and the predictive model

Points 1–4 will be illustrated in the following example, while point 5 will be dealt with in more detail in the next section. It should be noted that a model may not be satisfactory or new data may be available, and thus in reality the above is an iterative process, whereby models are always being updated and improved upon where possible.

7.2.1 Example 1: Job Time Prediction Under Varying Demand

The following data has been collected by a small manufacturing firm for 'peak demand', 'normal demand' and 'low demand'. The manufacturing workflow consists of three sequential stages (Fig. 7.4), each stage consisting of one machine. On arrival, jobs are put straight into a queue to be processed. Currently the firm provides a very rough estimate of completion time for a job (Table 7.1).

In order to get an initial idea of any potential relationships, it is worth plotting the data to see if any patterns emerge (Fig. 7.5). Immediately we see two obvious trends, **Job No** against **Arrival Time** and **Job No** against **Completion Time**; however with some thought this is an obvious relationship as the jobs are sequentially added to the system. In fact if the data was presented in an unordered manner we would not see

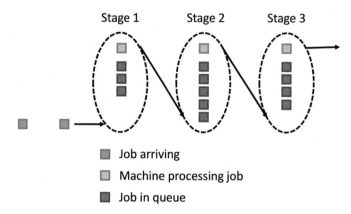

Fig. 7.4 Example 1 process

Table 7.1 Example 1 data collected at peak demand

Job no	Arrival time	Completion time	No jobs in system	Job time
1	0.99	21.52	0	20.53
2	1.48	29.61	1	28.13
3	6.60	37.53	2	30.94
4	12.56	45.49	3	32.93
5	15.40	53.32	4	37.92

this trend and this demonstrates the need to be cautious in identifying relationships between features.

Examining the plot of **No Jobs in System** (on arrival of job) against **Job Time** we see a strong linear relationship between the two features with which we can now build a predictive model.

7.2.1.1 Linear Regression Model

Figure 7.6 displays the plot for **No Jobs in System** (on job arrival) against **Job Time** accompanied by the linear regression model. Figure 7.7 displays the predictions for 50 test points that have not been included in the model for validation purposes and Fig. 7.8 plots the Absolute Percentage Error for each prediction. It is notable that prediction is poorer for the low demand case, which is sensible as increasing the jobs in the system causes it to act as a congested queue and therefore decreases the variability, making prediction easier.

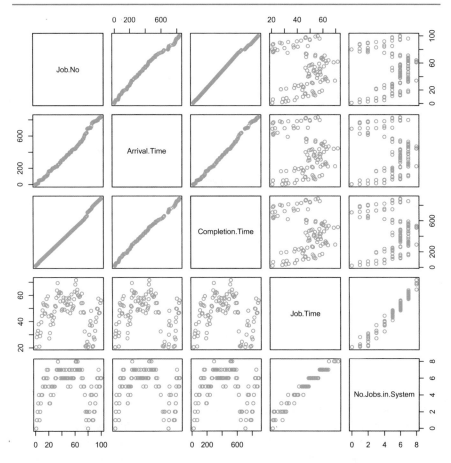

Fig. 7.5 Scatter matrix plot of peak data

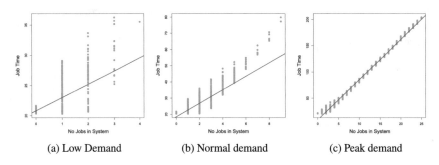

(a) Low Demand   (b) Normal demand   (c) Peak demand

Fig. 7.6 Number of jobs in the system 'versus' job time plots for low, normal and peak demand (blue), plotted with linear regression predictions (red)

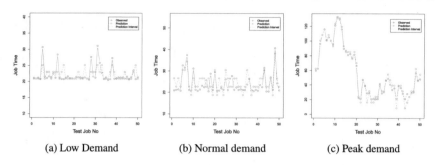

(a) Low Demand   (b) Normal demand   (c) Peak demand

Fig. 7.7 Plots of the observed and predicted job times from the test set

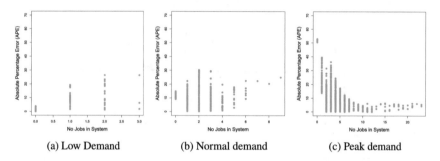

(a) Low Demand   (b) Normal demand   (c) Peak demand

Fig. 7.8 Absolute percentage error

7.2.1.2 K-Nearest Neighbours Model

For comparison we consider the K-Nearest Neighbours (KNN) model which predicts the target based on the mean value of the K-Nearest Neighbours. A fundamental difference between this model and the previous is the choice of the hyperparameter K. A hyperparameter is parameter that is set before we start the training process, and thus different values of the hyperparameter will result in different models. The following model uses a value of $K = 3$, accuracy results for both models are given in Table 7.2.

Table 7.2 Normalised RMSE cross-validation scores

	Low			Normal			Peak		
	RMSE	MAE	R^2	RMSE	MAE	R^2	RMSE	MAE	R^2
Linear regression	0.758	0.478	0.431	0.557	0.461	0.690	0.091	0.072	0.992
KNN	0.728	0.430	0.462	0.449	0.326	0.794	0.057	0.045	0.997

7.2.1.3 Some Notable Issues

Although both these models have some predictive power, it struggles to make predictions at low demand. Why is this? Apart from the aforementioned issue, factors that could contribute are:

- that we have picked the wrong model;
- that we simply haven't collected data on all the features that contribute to the time taken by a job;
- and/or we have done a bad job in training the model.

This leads us to the following questions and perhaps the most important part of building a machine learning model. How can we improve our model and measure its predictive power?

7.3 Assessing the Predictive Power of a Model

For a predictive model to be useful, it must not only perform well on the training data, but also on the test data, we call this process model validation. Without a good measure of how well a model fits the data and its potential predictive power, we would not be able to substantially improve our model. Therefore it is inherent that we have good accuracy measures and validation methods.

7.3.1 Root Mean Squared Error (RMSE)

The most commonly used measure as to how well a model fits a dataset after training is the Root Mean Squared Error (RMSE), which measures the average of the squares of the errors. For a model $\hat{y} = \hat{f}(x)$ and n predictions , let the error (known as a residual error) of prediction i be defined as the difference between the prediction $\hat{y}_i$ and the observed value y,

$$e_i = \hat{y}_i - y_i,$$

then the MSE is defined as follows,

$$MSE = \frac{1}{n} \sum_{i=1}^{n} e_i^2$$

and,

$$RMSE = \sqrt{MSE}.$$

The lower the RMSE, the better the potential fit; however the scale of the data must be taken into account as a residual error of $e_i = 10$ would mean different things for housing prices and for train prices, one would be a very low error, the other a high error. Note that the square root is necessary to scale the data given that we squared the residual errors e_i.

7.3.2 Mean Absolute Error (MAE)

$$MAE = \frac{1}{n} \sum_{i=1}^{n} |e_i|$$

7.3.3 Mean Absolute Percentage Error (MAPE)

The Absolute Percentage Error is defined as,

$$APE\ (\%) = \left| \frac{e_i}{y_i} \right| \times 100,$$

and the Mean Absolute Percentage Error (MAPE) is given by,

$$MAPE\ (\%) = \frac{1}{n} \sum_{i=1}^{n} \left(\left| \frac{e_i}{y_i} \right| \times 100 \right)$$

7.3.4 Coefficient of Determination (R^2)

Define the mean of the data as:

$$\overline{y} = \frac{1}{n} \sum_{i=1}^{n} y_i$$

$$RSS = \sum_i e_i^2$$

$$TSS = \sum_i (y_i - \overline{y})^2$$

$$R^2 = \frac{TSS - RSS}{TSS} = 1 - \frac{RSS}{TSS}$$

- TSS measures the variability in the data (variance)
- RSS measures the variability in the data compared with the model predictions
- 1−RSS/TSS measures the ratio of the variability in our model compared with the variability in the data.

7.3.5 Underfitting and Overfitting

All predictive models must balance the bias–variance tradeoff, and this is a balance between a model with high bias that generalises (underfits) and a model with high variance that is over sensitive to training data (overfits). Both fail to capture the relationships between features and target and have little predictive power. It is therefore essential to have methods and tools available to correctly provide a balance between these two. Figure 7.9 plots models of increasing complexity, as the complexity increases the model overfits the test data resulting in poorer performance.

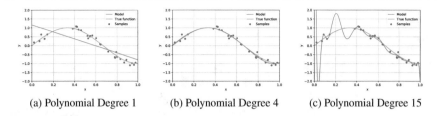

(a) Polynomial Degree 1   (b) Polynomial Degree 4   (c) Polynomial Degree 15

Fig. 7.9 Plots demonstrating underfitting and overfitting of a model

7.3.6   Cross-Validation

As previously noted, a key method in evaluating our model accuracy is to compare the predictions made on the test set with the observed values of the test set. However, we have two potential issues, first the test set may be too small to properly evaluate the accuracy of our model and second we have introduced some bias by splitting the data into a training and a test set; i.e. a different training/test set split might yield wildly different results. Cross-validation is a technique which helps overcome both of these issues and ensures that we get a model that performs well across all data that we have to hand, thus increasing the likelihood of good prediction on unseen data.

The basic principle of cross-validation is to use different splits of the dataset to train different versions of the model. Each of these is then scored for accuracy, and the overall accuracy of the model is given by the average of these scores. Many methods exist to perform cross-validation, but focus is given to the popular k-fold method.

7.3.6.1   K-Fold Cross-Validation

The idea behind k-fold cross-validation is to split the dataset into k non-overlapping groups with one group being used as the test (validation) set and the remaining data used to train the model. This is repeated so that each of the k groups is used as the test set. For example, a fivefold cross-validation would results in five groups and we would train and validate five models, each model using different groups for the training and test set. Figure 7.10 shows an example of this.

7.3.6.2   Cross-Validation Results for Example 1

Cross-validation is applied to the two models used in Example 1, and results are displayed in Table 7.3. Different accuracy scores appear for the five folds demonstrating the difference in model accuracy as the training/test dataset split is changed. The overall accuracy which is calculated as the mean of the fivefold scores is also displayed.

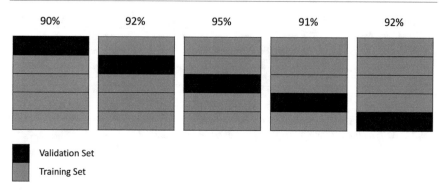

Fig. 7.10 K-fold validation for five folds, the model is tested using five groups to assess its accuracy

Table 7.3 Fivefold cross-validation RMSE scores for Example 1 (normal demand)

	1st Fold	2nd Fold	3rd Fold	4th Fold	5th Fold	Average
Linear regression	0.572	0.570	0.515	0.591	0.537	0.557
KNN	0.450	0.480	0.443	0.440	0.433	0.449

7.3.7   Learning Curves

Learning curves are an extremely useful in visualising the accuracy of prediction based on the increase in the amount of data used to train the model. A major benefit of this can be the time saved in realising that if the model fails to improve at a given point, then adding more data won't help (retraining some models can take hours, thus saving time can be extremely important). To further improve the model we might have to consider feature selection and/or different models, which are mentioned in the next section.

A learning curve for Example 1 is plotted in Fig. 7.11a for the KNN model, and we can conclude that our model is fairly stable after approximately 600 training examples.

7.3.8   Validation Curves

Validation curves are similar to learning curves; however they assess the accuracy of prediction based on the tuning of a hyperparameter. Figure 7.11b displays the validation curve for the KNN model given in Example 1. Clearly a value of $K = 3$ is close to optimal. It should be noted that many machine learning techniques have multiple hyperparameters, and tuning these for optimal performance can be tricky and time-consuming; however methods such as grid search automate this procedure to a degree.

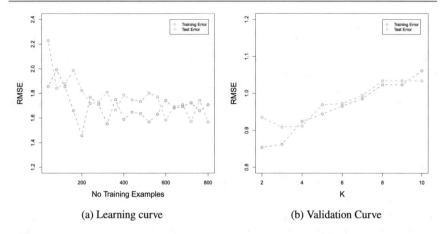

(a) Learning curve   (b) Validation Curve

Fig. 7.11 Plots demonstrating learning and validation curves for the KNN multiple regression model for Example 2

7.4   How to Improve Model Accuracy

Having established how we can measure and visualise the predictive accuracy of a model, we can now look at various methods that may lead to potential improvements to the model.

7.4.1   Feature Selection

Determining the features x that have a relationship with a given target y can be difficult, but is extremely important for model accuracy. Trial and error is by no means a bad idea, and in fact it can result in some fairly good models; however there are some techniques to help this.

7.4.1.1   Exploratory Plots
It is always useful to graph the data as this can reveal natural relationships between features. Figure 7.5 shows an example of this, and the relationship between **No Jobs in System** and **Job Time** was easily identifiable.

7.4.1.2   Filter Methods
Filter methods use some form of statistical analysis to score and then rank the various features. These methods tend to consider each feature independently, so it will not include information about relationships between multiple features and their effect on a target. While it can give you an idea of the important features to include in a model, it does not guarantee that this is the best combination.

7.4.1.3 Wrapper Methods

Wrapper methods are methods which search for the best combination of features. The basic idea is to pick a set of features, evaluate the model accuracy, then repeat this process and compare the results of the models. This can be done using many search methods; however it should be noted that given n possible features, the number of combinations is of size 2^n. Thus careful consideration needs to be given to the search method employed.

7.4.1.4 Embedded Methods

Embedded methods are machine learning techniques which incorporate feature selection into their learning process. That is they seek to eliminate some of the features which are not seen to contribute to the prediction, and therefore the model becomes simpler and may generalise better to unseen data. Examples of embedded methods are regularised regression methods such as Lasso and Ridge regression.

7.4.2 Example 2: Improving the Model with Additional Information (Multiple Regression)

Continuing with our three-stage process, we now have an additional (replica) machine available at each stage (Fig. 7.12). We are also provided with additional information of how many jobs are currently at a given stage (includes waiting and currently being processed jobs) (Table 7.4).

Applying the previous linear regression model shows that model accuracy has now decreased (the RMSE has increased) (Table 7.5). However we can now take advantage of the extra features that have been stored in the dataset. Postulating that the time taken for a job is related to the number of jobs at each of the three stages,

$$y = \beta_0 + \beta_1 x_1 + \beta_2 x_2 + \beta_3 x_3 + \epsilon,$$

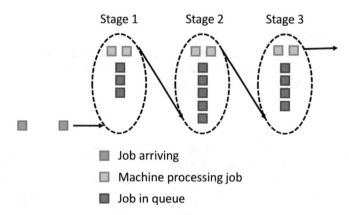

Fig. 7.12 Example 2 process

Table 7.4 Example 2 data collected at peak demand

Job no	Job time	No jobs at Stage 1	No jobs at Stage 2	No jobs at Stage 3
47	22.18	0	2	2
48	20.87	0	1	2
49	21.16	1	1	2
50	20.82	0	0	1
51	20.91	1	0	0

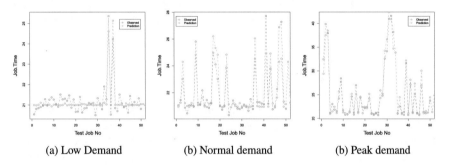

(a) Low Demand   (b) Normal demand   (b) Peak demand

Fig. 7.13 Plots of the observed and predicted job times given by the KNN multiple regression model for Example 2

we can train a multiple regression model of the form,

$$\hat{y} = \hat{\beta}_0 + \hat{\beta}_1 x_1 + \hat{\beta}_2 x_2 + \hat{\beta}_3 x_3.$$

Table 7.5 displays the RMSE cross-validation scores for the simpler regression models of the previous example and the more complicated multiple regression models. Figure 7.13 plots the predictions of 50 test data points for low, normal and peak demand. It is clear that the KNN (multiple) model is the most accurate in making predictions.

7.4.3   More Data

An increase in the amount of data to hand is generally a good thing, and it is worth collecting as much data as possible; however, does it improve our model if we increase the size of the dataset? The answer is problem specific and dependent on the dataset and complexity of the underlying relationships. In most situations though, there is a certain point where the accuracy of the model will no longer increase as more data is added. This can be analysed by plotting a learning curve as mentioned in the previous section.

Table 7.5 Fivefold cross-validation normalised RMSE scores for Example 2 (normal demand)

Model	1st fold	2nd fold	3rd fold	4th fold	5th fold	Average
Linear regression	0.667	0.666	0.717	0.733	0.692	0.695
KNN (Linear)	0.565	0.557	0.568	0.520	0.530	0.548
Multiple linear regression	0.533	0.555	0.474	0.622	0.507	0.538
KNN (Multiple)	0.350	0.340	0.319	0.336	0.282	0.326

7.4.4   Compare Models

All of the above can be applied to different machine learning models such as regression trees, polynomial regression and regressions splines to assess how well each of these techniques works on a specific dataset and problem. In general multiple techniques should always be applied to a problem to not only determine the best approach, but also get a feel for the dataset. Applying one technique might increase understanding, even if it does not directly replace the current best model. The following example illustrates the different accuracy's of three different techniques in modelling an extension of Examples 1 and 2.

7.4.5   Example 3: Multiple Job Types (Model Comparison)

The firm has expanded and has a total of two uniform machines per stage and is now required to make three different types of the same product (Fig. 7.14), with varying manufacturing sizes at each stage. The standard job has sizes (1, 1, 1) for each of the three stages; however the two new types of product have varying sizes per stage, (1, 1, 2) and (1, 2, 1). Therefore if a machine at Stage 1 requires approximately 6 min for the standard job, jobs 2 requires 12 min. The company makes approximately 50% of the standard product, 30% of product 2 and 20% of product 3.

It is obvious that the different jobs will make prediction harder, and we may wish to include this information in our model. We analyse three models, whose input features include:

- Model 1—{**No Jobs (S1), No Jobs (S2), No Jobs (S3)**}
- Model 2—{**Job Size (S1), Job Size (S2), Job Size (S3)**}
- Model 3—{**No Jobs (S1), No Jobs (S2), No Jobs (S3), Job Size (S1), Job Size (S2), Job Size (S3)**}

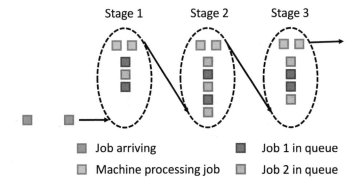

Fig. 7.14 Example 3 process

Table 7.6 Example 3 data collected at peak demand

Job no	Job time	No jobs at Stage 1	No jobs at stage 2	No jobs at stage 3	Stage 1 job size	Stage 2 job size	Stage 3 job size
100	29.13	0	0	0	1	1	2
103	21.39	0	1	2	1	1	1
102	31.24	1	1	0	1	1	2
104	29.23	1	1	2	1	1	1
105	24.51	2	0	3	1	1	1

To test the accuracy of these models we can compare them not only against each other, but against the previous model that did not include job size data (Table 7.6).

Accuracy results are presented in Table 7.7 for three different techniques which have been generated via cross-validation. Optimal values of the hyperparameters for KNN and Gradient Boosting were found using a restricted grid search. Note that Gradient Boosting has numerous hyperparameters and could be further tuned for increase accuracy.

It is clear from these results that for all three techniques, model 1 is poor in comparison with model 2 and model 3. Furthermore, model 3 proves to be a slight improvement on model 2.

7.5   Generating Data Via Simulations

Throughout this chapter we have not discussed how this data is collected, and we have assumed that we have the data and would like to construct predictive models. A simple method is to manually book keep; however this requires extra resource and may also prove not feasible due to the speed or complexity of parts of the production planning stage. Sensor capture enables a cheap automated way of capturing data to

Table 7.7 Fivefold cross-validation RMSE scores for Example 3 models

	Average CV score								
	Model 1			Model 2			Model 3		
	Low	Normal	Peak	Low	Normal	Peak	Low	Normal	Peak
Multiple linear re-gression	0.998	0.970	0.909	0.178	0.441	0.598	0.168	0.366	0.447
KNN	1.001	0.968	0.911	0.184	0.437	0.597	0.187	0.375	0.451
Gradient boosting	1.011	0.997	0.895	0.243	0.458	0.578	0.114	0.271	0.419

use in building predictive models, and it is easy to imagine data being collected and progressively building and improving models to enhance the predictive information a firm has to hand.

What though does one do in the absence of data? There are resampling methods such as bootstrapping, where additional data points can be generated by sampling from the existing data with replacement. However, this still requires data to begin and does not take into account the absence of certain data, for example, only job times have been recorded.

An alternative approach is to build a simulation model from the known information such as machine processing times and average job arrival times—in fact the data in the prior examples was generated in a similar manner. This can be a very useful technique in scenario-based planning, allowing a cheap way to explored different situations and changes to a manufacturing process. The following example explores how a simulation can be affected by the uncertainty of machine processing times.

7.5.1   Example 4: Simulating Data Under Uncertainty

Here we set up a situation where machine processing times are uncertain, and different hypotheses of the machine processing times are used in the simulations that are built. The simulated data is then used to build a predictive model which is then used to make predictions on a ground truth simulation where machine processing times are distributed normally; for example a certain machine might have a mean of 6.5 and a standard deviation of 0.125.

From collected data about the machine (or worker estimates), we are provided with a minimum time (e.g. 6.25), a maximum time (e.g. 6.75). We estimate the average (mean) time to be the midpoint (e.g. 6.5). Using this information the following hypotheses are proposed:

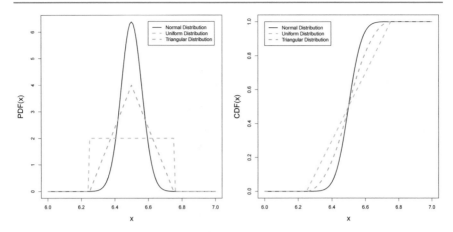

Fig. 7.15 PDF and CDF plots for estimate distributions

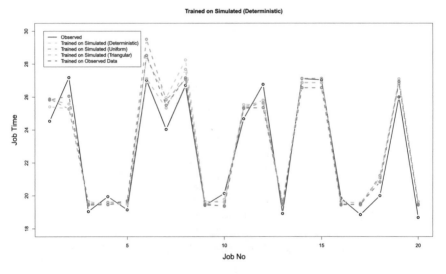

Fig. 7.16 Predictions given by the different machine distribution hypotheses

- Machine times are deterministic; i.e. machine times are given by the average machine time.
- Machine times are stochastic and estimated by sampling from a uniform distribution (min and max machine times required).
- Machine times are stochastic and estimated by sampling from a triangular distribution (min, average and max machine times required).

Figure 7.15 plots the uniform and triangular estimates of the unknown normal distribution. Table 7.8 and Fig. 7.16 display the results of predictions given by the models trained using data generated by the simulated hypotheses.

Table 7.8 Fivefold cross-validation RMSE scores for simulation prediction models

	Average CV score											
	Average time			Uniform distribution			Triangular distribution			Actual data		
	Low	Normal	Peak	Low	Normal	Peak	Low	Normal	Peak	Low	Normal	Peak
KNN (Multiple)	0.380	0.412	0.497	0.372	0.401	0.483	0.365	0.394	0.487	0.347	0.357	0.394

It is interesting to note that the average time approximation performs as well and in some cases better than the more complicated distributions. Furthermore, unless we have a large amount of data about a given machine, using a complicated distribution may in fact be further from the truth and add in unnecessary variability into the simulation. Thus average machine time may prove in many cases to be good enough to produce good simulations.

7.5.2 Kernel Density Estimation and Sampling

The previous hypotheses where a relatively crude attempt to estimate the distribution. However, we can do better using a technique called Kernel Density Estimation (KDE). We can then use this estimate of the distribution to draw samples from for our simulation. Figure 7.17a plots the unknown (real) distribution and the KDE estimate to illustrate how the accuracy of the estimate increases as we collect more data. This would not only provide a good estimate of the underlying distribution of the processing time, but also allow one to maintain a current distribution, by only considering the last n machine times (e.g. $n = 50$). As a example we simulate times from a machine that suddenly slows by a certain amount of time. 105 data points (original faster state of the machine) were generated from a normal with mean of 360.5 and standard deviation of 0.5 and 45 data points (current slower state of the machine) from a normal with mean of 370 and standard deviation of 0.7. Figure 7.17b shows a KDE estimate using all data points, and three KDE estimates using 20 sequential data points 1–20, 101–120 and 131–150. Clearly we have very different estimates of the current state of the distribution governing machine processing times; however the last plot (131–150) removes the bias of the old data points and only considers the current behaviour of the machine, thus replicating the new slower state of the machine. The choice of the number of data points to maintain is of course problem and domain specific and would require some careful analysis.

7.6 Worked Examples in R

Worked examples of the models that have been explored in this chapter are given as code extracts in the R language, and the datasets can be downloaded from https://bit.ly/3DUmD12.

7.6.1 Linear Regression Model

The first step to building a model is to import the dataset. Once imported we can view the column names, attach the column names to the workspace (we can refer to colname instead of dataset$colname) and plot a scatter matrix to look for potential relationships between features.

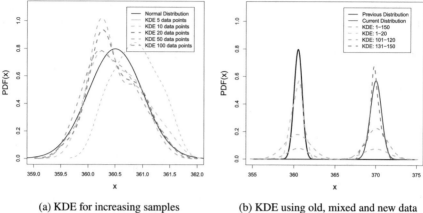

(a) KDE for increasing samples　　　　　　(b) KDE using old, mixed and new data

Fig. 7.17 KDE plots

We will also make use of the `train` method from the `caret` library for training our models.

```
# load caret library
library(caret)

# import dataset
dataset = read.csv("ex_1_peak_demand.csv", header=TRUE)

# list all the column names in the dataset
names(dataset)

# attach the names to the environment
attach(dataset)

# plot Scatterplot Matrix for selected columns
pairs(~Job.No+Arrival.Time+Completion.Time+Job.Time+No.Jobs.in.
    System, col="blue")
```

After importing the data we can split the data into a training set and a test set. Here we use an 80%/20% split

```
# set the random seed to replicate results
set.seed(17)

# randomly select 80% of the dataset
trainrows = sort(sample(nrow(dataset), nrow(dataset)*.8))

# create training dataset
train<-dataset[trainrows,]

# create test dataset
test<-dataset[-trainrows,]
```

We can now train a linear regression model `Y~X`, in the case of this dataset we consider `Job.Time~No.Jobs.in.System`.

```
# define the model training data
train.X = train["No.Jobs.in.System"]
train.Y = train$Job.Time

# define the model test data
test.X = test["No.Jobs.in.System"]
test.Y = test$Job.Time

# create linear model y~x
lm.fit = train(train.X, train.Y, method="lm")

# view summary statistics for the model, e.g. RMSE, R^2
summary(lm.fit)
```

Having trained our linear model we can visualise this on the training set as follows.

```
# plot the dataset, x = No.Jobs.in.System, y = Job.Time
plot(train.X, train.Y, col="blue", xlab="No Jobs in System", ylab=
    "Job Time")

# add the linear regression model to the dataset plot
abline(lm.fit, lwd =3, col="red")
```

With our trained model, we can also make predictions on the test dataset, evaluate these with RMSE and plot some predictions and observed values to consider the accuracy of the model

```
# predict Job Times based on linear model
pred.Y = predict(lm.fit, test.X)

# calculate RSME
print(sqrt(mean((pred.Y-test.Y)^2)))

# Plot model predictions for points 1 - 20
plot(test.Y[1:20], xlab="Test Job No", ylab="Job Time", col="blue"
    , type="b", ylim=c(0,130))

# Plot observed values for points 1 - 20
points(pred.Y[1:20], col="orange", type="b")
```

We can also build a simple KNN regression model for the same relationship Job.Time~No.Jobs.in.System. The train method automatically tries values of $k = \{3, 5, 7\}$, an example of how to change these values is given later on.

```
# train a knn regression model with neighbours
knnreg.fit = train(train.X, train.Y, method="knn")

# perform prediction for test dataset
pred.Y = predict(knnreg.fit, test.X)

# calculate RSME
print(sqrt(mean((pred.Y-test.Y)^2)))

# Plot model predictions for points 1 - 20
plot(test.Y[1:20], xlab="Test Job No", ylab="Job Time", col="blue"
    , type="b", ylim=c(0,130))

# Plot observed values for points 1 - 20
points(pred.Y[1:20], col="orange", type="b")
```

7.6.2 Multiple Regression Model

If we do not have a linear relationship in one variable, then it is sensible to consider a multiple regression model. Here we train a KNN regression model for the more complicated dataset used in Example 3.

First we must import a different dataset and set up the training/test data and column names used in the model.

```
# load the caret library
library(caret)

# import dataset
dataset = read.csv("ex_3_normal_demand.csv")

# attach the dataset names
attach(dataset)

# set the random seed to replicate results
set.seed(23)

# randomly select 80% of the dataset
trainrows = sort(sample(nrow(dataset), nrow(dataset)*.8))

#creating training dataset
train<-dataset[trainrows,]

# define the column names for X
colnames= c("No.Jobs.at.Stage.1", "No.Jobs.at.Stage.2", "No.Jobs.
    at.Stage.3", "Stage.1.Job.Size", "Stage.2.Job.Size", "Stage.3.
    Job.Size")


# define the training and test sets
train.X = train[,colnames]
train.Y = train$Job.Time

# define the training and test sets
test.X = test[,colnames]
test.Y = test$Job.Time
```

Now we can train the model using the train method from the caret library.

```
# fit a knn regression model
knnreg.fit = train(train.X, train.Y, method = "knn")

# view the results for the training of the model
knnreg.fit$results

# view the best value of k for the final model
knnreg.fit$bestTune

# perform prediction for test dataset
pred.Y = predict(knnreg.fit, test.X)

# calculate RSME
print(sqrt(mean((pred.Y-test.Y)^2)))

# Plot observed values for points 1 - 20
plot(test.Y[1:20], xlab="Test Job No", ylab="Job Time", col="blue"
    , type="b", ylim=c(0,50))

# Plot model predictions for points 1 - 20
points(pred.Y[1:20], col="orange", type="b")
```

Our model outputs a prediction RMSE of 1.1496 using a value of $K = 5$. It is also worth noting that the `train` method only considered values of $K = 5, 7, 9$

7.6.3 Cross-Validation

The previous model can be further validated and tuned by using the `trControl` and `tuneGrid` parameters for the `train` method. Here we will use a fivefold cross-validation and consider values of K between 1 and 10

```
# define the control parameters for training
train.control = trainControl(method = "cv", number = 5)

# define the values of the hyperparameter K to consider
tune.grid = expand.grid(k=1:10)

# fit a knn regression model
knnreg.cv.fit = train(train.X, train.Y, method = "knn", trControl=
    train.control, tuneGrid=tune.grid)

# view the results for the training of the model
knnreg.cv.fit$results

# view the best value of k for the final model
knnreg.cv.fit$bestTune

# perform prediction for test dataset
pred.Y = predict(knnreg.cv.fit, test.X)

# calculate RSME
print(sqrt(mean((pred.Y-test.Y)^2)))
```

Our model now selects a K value of 1 and results in a prediction RMSE of 1.283. A slight improvement on the previous model.

7.6.4   KDE Estimation of Distribution

To demonstrate how to estimate an unknown distribution from a given number of samples, 50 samples have been drawn from the normal distribution used in Sect. 7.5.2. Figure 7.18a plots the output of the following code:

```
# define x values (used in plot for comparison)
x <- seq(355,375, by=.01)

# calculate normal distribution (used in plot for comparison)
y = dnorm(x, 360.5, 0.5)

# Sample 50 points from normal distribution
x.samp = rnorm(50, 360.5, 0.5)

# Plot normal distribution and KDE estimates for 5,20 and 50
    samples
plot(x, y, col="black", lty=1, lwd=2, xlim=c(359,362), ylim=c(0,
    1.1), type="l", ylab="PDF(x)", cex.lab=1.5, main="KDE Estimate
    of Normal Distribution")
lines(density(x.samp[1:5]), lty=2, lwd=3, col="orange")
lines(density(x.samp[1:20]), lty=2, lwd=3, col="blue")
lines(density(x.samp), lty=2, lwd=3, col="red")
legend(359, 1.1, legend=c("Normal Distribution", "KDE 5 data
    points","KDE 20 data points", "KDE 50 data points"),
        col=c("black", "orange", "blue", "red"), lty=c(1,2,2,2),
            lwd=c(2,3,3,3))
```

As mentioned in Sect. 7.5.2, we can also sample from this distribution. To verify that this is the case, we have drawn 5000 samples from the KDE estimate and then used another KDE estimate on these samples for comparison. Figure 7.18b plots the results of this experiment.

```
# store KDE estimate for normal distribution
kde = density(x.samp, n = 2^10)

# sample from estimate
kern.samp = sample(kde$x, 5000, replace=TRUE, prob=dens$y)

# plot KDE estimate for normal distribution
plot(density(x.samp), lty=1, lwd=2, main="Sampling From KDE
    Estimate", ylim=c(0,1.1))

#plot KDE estimate of KDE estimate to verify samples are similar
lines(density(kern.samp), lty=2, lwd=3, col="red")
legend(359.1, 1.1, legend=c("KDE Estimate Distribution", "KDE
    Estimate From Samples"),
        col=c("black", "red"), lty=c(1,2), lwd=c(2,3))
```

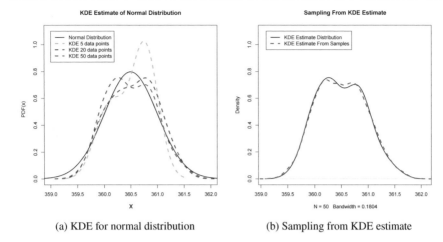

(a) KDE for normal distribution (b) Sampling from KDE estimate

Fig. 7.18 KDE plots from sample code

References

1. Syntetos AA, Boylan JE, Disney SM (2009) Forecasting for inventory planning: a 50-year review. J Oper. Res. Soc. 60(sup1):149–160. https://doi.org/10.1057/jors.2008.173
2. Zanjani MK, Nourelfath M, Ait-Kadi D (2010) A multi-stage stochastic programming approach for production planning with uncertainty in the quality of raw materials and demand. Int J Prod Res 48(16):4701–4723. https://doi.org/10.1080/00207540903055727
3. Mula J, Peidro D, DÃnaz-MadroÃśero M, Vicens E (2010) Mathematical programming models for supply chain production and transport planning. Euro J Oper Res 204(3):377–390. https://doi.org/10.1016/j.ejor.2009.09.008
4. Asgari N, Nikbakhsh E, Hill A, Farahani RZ (2016) Supply chain management 1982–2015: a review. IMA J Manag Math 27(3):353–379
5. Chaudhary K, Singh M, Tarar S, Chauhan DK, Srivastava VM (2018) Machine learning based adaptive framework for logistic planning in industry 4.0. In: Singh M, Gupta PK, Tyagi V, Flusser J, Ören T (eds) Advances in computing and data sciences. Springer, Singapore, pp 431–438
6. Al-Aqrabi H, Liu L, Hill R, Antonopoulos N (2012) Taking the business intelligence to the clouds. In: 2012 IEEE 14th international conference on high performance computing and communication; 2012 IEEE 9th international conference on embedded software and systems. IEEE, pp. 953–958

Part III
Application

Part III contains further explanations of the use of analytics techniques in an industrial scenario. Additionally, a study on the adoption of analytics technologies in the UK illustrates some of the critical success factors for digital disruption within industry.

Case Study: Confectionery Production

8

8.1 Introduction

A confectionery production firm produces sugar- and chocolate-based confectionery mainly sold through its own retail outlets. Difficulties with the flow and exchange of timely information made the allocation of production resources challenging, resulting in poor operating and financial performance.

The company's operating objectives can be summarised as:

- Production—manufacture a full range of confectionary so that their retail outlets need only stock their own products;
- Sales—to satisfy all (almost all) sales demand by carrying sufficient stock to satisfy the expected demand;
- Quality—to produce goods of a noticeably higher quality than those produced by competing firms;
- Price—where comparable the unit prices should be the same as those charged by competing manufacturers.

The objective of the study was to provide the management of the company with a means of scheduling production in a department where several different types of product were manufactured. A subsequent objective was the identification of the dataset necessary to enable efficient and effective planning and control.

These management requirements were satisfied in two stages:

- The first stage being the determination, on a monthly basis, of the production time, in days production, to allocate to the manufacture of each product line where this allocation will be subject to machinery and manning restrictions;
- The second stage involving the creation of a full set of acceptable daily production pairs.

© Springer Nature Switzerland AG 2021
R. Hill and S. Berry, *Guide to Industrial Analytics*, Texts in Computer Science,
https://doi.org/10.1007/978-3-030-79104-9_8

Concurrent with this work was an analysis of the production unit with the aim of determining the machinery and manpower potential and availability.

To satisfy the first aim several possible methods of solution were examined, with particular reference to the special nature of the company. Typical problems were modelled and solved to determine the most efficient and effective approach.

8.1.1 Company Organisation

K and C Fletcher operates as two distinct divisions production and sales. Problems considered here are concerned with Hard Boiled (HB) Confectionery Production and Planning HB Production.

8.1.2 Production

Fletcher's factory operates as four independent production units each being responsible for a particular set of lines. This case study is based around one of these units, the HB confectionary unit.

Production in this unit requires more skilled labour than the other production units in the factory, and the production processes are more labour intensive than those in the other production units; therefore training given to new employees is lengthy and staff retention is critical to the efficient operation of this unit.

Planning problems for all units were caused by:

- Storage restrictions, this results from the allowed shelf life for products and the limited size of the available cold storage facilities. A lesser problem for the HB unit as these products tend to have long shelf lives;
- Satisfying demand, demand for the firms produce is highly seasonal, and weather dependent, particularly for the confectionary produced by this unit for which there are many alternative suppliers.
- Packaging, the tendency of the company to develop lines with an individual image will require special distinctive packaging;
- Poor communication between retail outlets, planning office, production and inventory (Fig. 8.1);
- Stock level data was updated on a monthly basis, and there was little (immediate) communication between these functions.

The company had employed a Taylor-based philosophy, work study, when setting daily production targets. Unfortunately, within this planning they considering each line in isolation not being able, through a lack of information, to assess the feasibility of these (production) quantities when two lines were produced simultaneously, standard unit working practice, thereby causing the planning department to set impossible production targets.

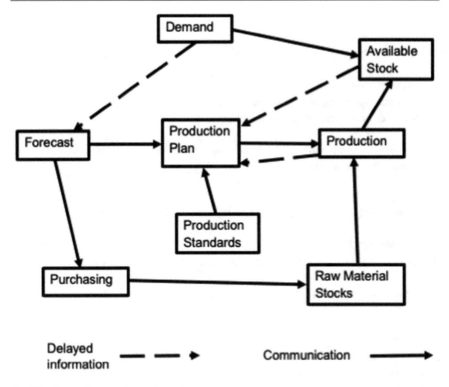

Fig. 8.1 Flow of information and goods

The retail outlets reported sales on a regular basis. These values were then aggregated and input into the planning process; thus the forecast was based on a hypothetical 'average retail outlet'.

The HB confectionery unit was chosen for this case study because the company had observed that it had been operating inefficiently in comparison with the other production units within the factory.

They had observed that:

- The unit was not achieving its planned daily production;
- There was a lack of stock on many occasions during the periods of peak demand.

The immediate needs, of the investigation, seemed to be:

- Develop/derive a long-term scheduling system based around the current need for a monthly production plan;
- Develop/derive a short-term daily scheduling system.

The current problems were thought to have been the result of problems experienced in one or more of the areas:

- Forecasting demand;
- Setting production plans;
- Lack of packaging;
- Lack of appropriate production machinery;
- Staff shortages;
- Inefficient operation of the production unit;
- Lack of communication within the company.

The symptoms displayed by this unit suggested the sources of these problems: Low efficiency, caused by:

1. Poor plan;
2. Insufficient or inappropriate machinery;
3. Low morale;
4. Staff shortages;
5. Poor communication between planning sales and manufacturing;

(Three and four following from points 1 and 2) and insufficient stock levels, caused by:

1. Poor forecast;
2. Poor production plan;
3. Low efficiency during production;
4. Staffing problems;
5. Poor communication between planning sales and manufacturing.

The case study then proceeded through an examination of current planning systems, and the subsequent development of improved planning systems, and an evaluation of the resources present within this unit to demonstrate that the low productivity reported in this unit is a direct consequence of the department operating with insufficient/inappropriate machinery.

The company would not allow any investigation into improvements to the forecasting function employed by the company. Currently the forecast sales over the next year were obtained through the addition of a fixed percentage to the current sales.

8.2 Hard Boiled Confectionery Organisation and Planning

The unit produces through the year quantities of approximately 30 different items subdivided into three product types by final cut/wrap process (Fig. 8.2). There are the following constraints to be considered by the planners when setting a monthly (4 weekly) production plan:

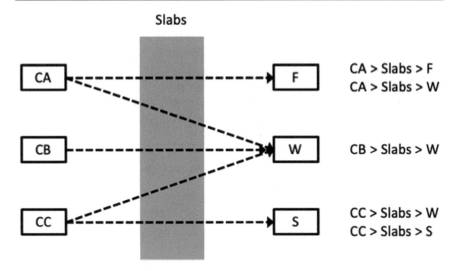

Fig. 8.2 Production routings

- These items have a limited shelf life, albeit long.
- This is seasonality in the demands for these products, the peak demand mainly coinciding with the factory holiday period.
- Production is to be planned in full working days, to reduce lost time due to between lines clean up between products.

On any day only two of the possible production type of product can be made; see Fig. 8.2. Generally exactly two items are produced every day.

8.2.1   Unit Operation

All lines produced in this department will pass through three production stages:

1. Cooking (CA or CB or CC)
2. Make up (5 identical slabs)
3. Cut and wrap (F or W or S)

The layout is shown in Fig. 8.2, and the possible product flows are indicated by the dashed arrowed lines.

A typical progression of raw materials into finished goods can be illustrated through a description of the processes involved in the manufacture of a wrapped 'rock' line:

1. Cooking stage, here the raw materials (sugar, etc.) are transformed into their working state, a hop syrup
2. Make up, here colouring and flavouring are added to the syrup and the mixture allowed to set into a semi-solid form so that it could be worked into a suitable shape (the pattern in rock is constructed at this stage)
3. Cut wrap, the semi-set syrup is rolled into a conical mass which is moulded into the required cross section then sliced and wrapped by the ?whirl? machine (machine W).

The wrapped confection is then left in trays to cool the packed in boxed to be transferred to the onsite storage area.

The material flows for the three product types are:

- *Forplast*—CA, Slabs (at least two of the five slabs required);
- *Whirl*—CA, Slabs, W;
- *Square*—CC, Slabs, S.

As a consequence of machine availability it follows that on any one day no more than two lines can be produced simultaneously.

Hence during a normal working month (20 days), there are exactly 40 line days of production to be planned. Any extra production will require overtime working.

8.2.2 Scheduling

The firm produces long-term forecasts determining the number of working days required each month over a 12 month period, medium-term forecasts determining the quantity of each lines to be produced each month, and the short-term forecasts establishing the daily production pairs. In each case the company aims to satisfy demand while minimising costs, production plus holding.

Table 8.1 shows the number of days allocated to each production group by the firms planning department, notice that they have chosen to allocate some production to overtime working in month 7 month 10 although there had been sufficient production time in earlier months, a consequence of a lack of accurate information; see Fig. 8.3.

8.2.3 Model to Determine Optimal Long Term, Monthly Production Plans

This can be described as a linear programming model where:

- X_i is the required number of days production in month i;
- D_i is the demand expressed as the number of days production in month i;
- S_i stock held in store at the end of the month;

Table 8.1 Example production schedule showing the production days allocated by the planners

Line type	Product group	Staffing	Days allocated to this group in each month number									
			1	2	3	4	5	6	7	8	9	10
	W1	12	4	4	1	4	5	4	8	7	3	11
W	W2	11	5	4	1	3	3	3	2	2	3	1
	W3	?10	11	10	8	10	9	11	10	10	9	7
	Total Days/Month		20	18	10	17	17	18	20	19	15	19
	F1	13	6	4	3	4	8	4	4	5	3	7
F	F2	11	3	3	1	5	2	3	3	2	3	1
	F3	?10	3	6	3	2	6	4	8	2	7	8
	Total Days/Month		12	13	7	11	16	10	15	9	13	16
S	S	?10	8	9	3	10	7	7	6	12	4	7
Allocated production time			40	40	20	38	40	35	41	40	32	42
Available production time			40	40	20	40	40	35	40	40	40	40

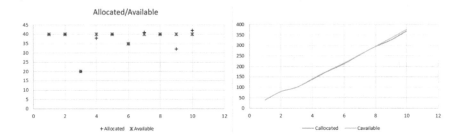

Fig. 8.3 Comparing allocated and available production time

- P_i the production cost of one days production;
- H_i the cost of holding a day's production in stock.

The best production plan can be obtained as the solution to the problem determine the production quantities Xi so that:

$$X_1 + S_0 - S_1 = D_1$$
$$X_2 + S_1 - S_2 = D_2$$
$$X_i + S_{i-1} - S_i = D_i$$

While minimising:

$$\sum P_i X_i + \sum H_i S_i$$

However, because:

$$\sum X_i = \text{Total demand}$$

Table 8.2 Shortage of production time in month 6

Month	1	2	3	4	5	6	7	8	9
Demand/ month	34	35	33	38	36	**35**	40	23	16
Available time	40	40	36	40	38	**30**	40	40	40
Shortage of time						**5**			

Table 8.3 Shortage of production time in month 5

Month	1	2	3	4	5	6	7	8	9
Demand/ month	34	35	33	38	**41**	30	40	23	16
Available time	40	40	36	40	**38**	30	40	40	40
Shortage of time					**3**				

is constant, given that the company aims to satisfy the forecast demand $P_i = P$ and $H_i = H$ the model can be simplified to become:

$$\text{Minimise } (\Sigma S_i)$$

$$X_1 - S1 = D_1$$

$$X_i + S_{i-1} - S_i = D_i \; i > 1$$

Indicating that the optimal production plan results from producing goods as late as possible (minimising stock holding costs).

An example of this is Fig. 8.4, showing the progression from forecast to production plan.

Plot 8.4a shows the demand plotted against available capacity.

Plot 8.4b indicates that there exists a production plan such that overtime working will not be required.

Plot 8.4c shows the derived production plan, and plot 8.4d shows the implied stock holding in terms of months held in stock.

To illustrate the derivation of an optimal production plan consider a problem where there is a shortage of production time (5 days) to satisfy the forecast demand during period 6 (Table 8.2).

Move the excess demand into month 5. This move has created an excess demand of 3 in period 5 as in Table 8.3.

Move the excess into month 4. This move has created an excess demand of 1 in period 4 (Table 8.4).

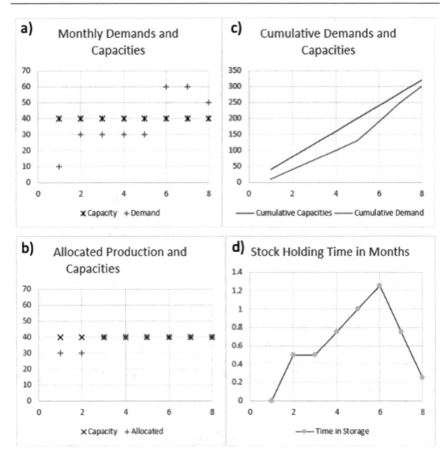

Fig. 8.4 Demand, production and stock holding plots

Table 8.4 Shortage of production time in month 4

Month	1	2	3	4	5	6	7	8	9
Demand/ month	34	35	33	**41**	38	30	40	23	16
Available time	40	40	36	**40**	38	30	40	40	40
Shortage of time				**1**					

Table 8.5 Feasible production plan

Month	1	2	3	4	5	6	7	8	9
Demand/ month	34	35	34	40	38	30	40	23	16
Available time	40	40	36	40	38	30	40	40	40
Shortage of time									

This move created an excess demand of 1 in period 4. Finally giving a feasible production plan by moving an excess of 1 into period 3 as per Table 8.5.

This procedure giving the optimal (minimum stock holding) cost solution. If the company now wishes to increase the quantity of stock held in their warehouse they can pull production forward into periods 1, 2 or 3.

Using a similar approach the medium term, weekly, production plans can be constructed by ensuring, when no overtime has been needed, that each product group is not allocated more production time than there are production days in the week.

Plot 8.5a, b, c, d illustrates the approach when overtime working is used to enable the company to satisfy the expected demand.

Here the required overtime working will be allocated as late as possible, to reduce stock holding costs; see Plot 8.5b, and this plot indicates that overtime working will be required during month 8.

8.2.4 Implementation of Monthly Planning

Table 8.6 lists the plans generated (by adopting a minimise stock holding cost approach) on a monthly basis over a 5 month period.

Table 8.6 Convergence of monthly production requirements. *Note* * and ** indicate periods that include public holidays and therefore reduced capacity

Forecast made in month	Month Number; each month containing 4 weeks, (at most 20 days), Production days allocated in each production plan									
	1	2	3*	4	5	6**	7	8**	9	10
1	40	40	20	40	40	36	41	40	32	42
2		40	20	38	40	36	40	38	30	40
3			25	39	40	36	40	38	30	40
4				40	40	36	40	38	30	40
5					40	36	40	38	30	40

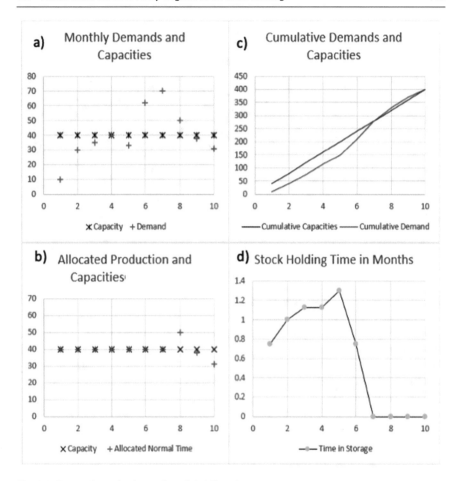

Fig. 8.5 Demand, production and stock holding plots

This table shows how this approach stabilises the derived plans, and the results from months 5 to 10 show how the required production time converges. The major divergence occurs as a result of the shortage of shipping capacity in month 3 when the factory concentrated on dispatching Christmas goods resulting in an apparent lack of hard boiled sugar confection in the retail outlets.

For a second comparison the month-end stock levels forecast at each month were collated and are shown with the actual stock levels in Table 8.7.

The overestimated month-end stock levels have resulted from seasonal shipping patterns and the selection of inefficient production pairs resulting in actual production quantities being less than the planned quantities.

Table 8.7 Comparing actual with estimated month-end stock

Forecasts	Month number each month containing 4 weeks, at most 20 days									
Made in month	1	2	3*	4	5	6**	7	8**	9	10
1	43.6	50.2	41.1	53.3	55.6	59.7	64.5	64.7	57.8	67.1
2		51.0	41.0	52.3	55.9	62.4	68.3	69.4	58.8	64.9
3			38.4	58.9	64.8	71.6	74.2	78.3	65.5	73.0
4				50.6	54.6	61.9	65.0	68.7	58.7	63.6
5					58.7	61.3	65.4	69.7	59.2	63.9
Actual stock	41.7	48.9	38.5	50.7	54.4					
Forecast less franchise	42.9	49.4	37.5	49.8	57.9					
Percent errors	1.9	5.1	2.6	1.8	6.4					

8.2.5 Allocating Production Pairs

The final stage, the construction of a set of daily production pairs was, initially, the more complex task given that these pairs were to chosen so that there will be an efficient production, satisfying production targets. This schedule is constructed by selecting daily pairs from the table of possible pairs.

From Table 8.8:

- An efficient pair is indicated by *
- An impossible pairing by *I*
- A possible pairing, low efficiency *P*

Notice that there are:

- Possible Pairs 191
- Efficient Pairs 98, (51%)
- Inefficient Pairs 67, (35%)
- Impossible Pairs 26, (14%)

This table showing that many production pairs were either impossible or must lead to an inefficient production plan. That is if the production planners had produced a daily plan consisting of an inefficient pairing then the workshop would have been prevented from demonstrating an 'efficient' operation using the firm defined criteria.

Such pairing had been (often) requested by the planners, and hence the workshop was not able to produce the 'established' quantity of goods and the resultant data would (unfairly) suggest, to the firms management, that this department is operating inefficiently!.

Table 8.8 Viability of production line pairings

	F1	F2	F3	F4	F5	F6	F7	S1	S2	S3	S4	S5	S6	S7	S8
W1	P	P	P	I	*	P	I	*	P	*	*	*	P	*	*
W2	P	P	P	P	*	P	P	*	P	*	*	*	P	*	*
W3	P	P	P	P	*	P	P	*	P	*	*	*	P	*	*
W4	*	*	*	*	*	*	P	I	*	I	I	I	*	I	I
W5	*	*	I	*	I	*	I	*	*	*	I	*	*	*	*
W6	*	*	I	*	I	*	I	P	*	P	I	P	*	P	P
W7	P	P	P	I	*	P	I	*	P	*	*	*	P	*	*
W8	*	*	I	*	I	*	I	P		P	I	P	*	P	P
W9	*	*	I	*	I	*	I	*	*	*	I	*	*	*	*
S1	*	*	P	*	P	*	P								
S2	P	P	P	P	*	P	P								
S3	*	*	P	*	*	*	P								
S4	*	*	P	*	P	*	P								
S5	*	*	P	*	*	*	P								
S6	P	P	P	P	*	P	P								
S7	*	*	P	*	P	*	P								
S8	*	*	P	*	*	*	P								

Inefficient pairings were requested within each months production plan, for example. A typical months requirement required pairing to satisfy the line demands:

- 20 production days for items of type W
- 10 production days for items of type F
- 10 production days for items of type S

The daily pairings constructed from these items were such that 25% of the pairings were inefficient, as defined in Table 8.8. Consequently the actual production attained will be less than the 'standard' expected quantity. This factor explained the unhappiness of the workforce who worked as efficiently as possible but were blamed for underachievement when the problem was caused by a lack of suitable machinery within the department.

Notice that were quantities of item $F7$ required as there is no efficient pair for this line then production must be inefficient.

Notice also that within this firm the production requirements for lines type W are generally produced during every working day. Therefore Table 8.9 only shows the feasibilities of combining a W line with either an S line or an F line.

- Possible Pairs 135
- Efficient Pairs 70, (52%)

Table 8.9 Viability when a W line must be chosen

	F1	F2	F3	F4	F5	F6	F7	S1	S2	S3	S4	S5	S6	S7	S8
W1	P	P	P	I	*	P	I	*	P	*	*	*	P	*	*
W2	P	P	P	P	*	P	P	*	P	*	*	*	P	*	*
W3	P	P	P	P	*	P	P	*	P	*	*	*	P	*	*
W4	*	*	*	*	*	*	P	I	*	I	I	I	*	I	I
W5	*	*	I	*	I	*	I	*	*	*	I	*	*	*	*
W6	*	*	I	*	I	*	I	P	*	P	I	P	*	P	P
W7	P	P	P	I	*	P	I	*	P	*	*	*	P	*	*
W8	*	*	I	*	I	*	I	P		P	I	P	*	P	P
W9	*	*	I	*	I	*	I	*	*	*	I	*	*	*	*

Table 8.10 Generated pairings

	F1	F2	F4	F6	F8	F9	F10	S1	S3	S4	S5	S6	S7
W1		1		1					1			1	
W2	2												
W4					1		1						
W5			2			1							
W6						1			2			2	1
W7							1			2			

- Inefficient Pairs 39, (29%)
- Impossible Pairs 26, (19%)

Table 8.10 shows the product pairs chosen during a typical month (some pairs occurring more than once), and Table 8.11 shows that this choice includes 6 inefficient production pairs thus demonstrating that the implied production plan will not achieve its target production (this plan used several inefficient pairings).

Notice that were quantities of item $F7$ required as there is no efficient pair for this line then production must be inefficient.

The best set of production pairs, for this dataset, can be generated using a linear programming model using as an example a schedule where the demand for each line type was:

- W = 20, S = 10 and F = 10 then if
- $WF_{ij} = 1$ lines W_i and F_j have been chosen as a production pair
- $WS_{ij} = 1$ lines W_i and S_j have been chosen as a production pair
- RWF_{ij}=0, or 1, or 99
- RWS_{ij}=0, or 1, or 99; 0 implies efficient, 1 inefficient, 99 impossible
- WD, FD, SD are the demands, in working days for items of type W, F and D

Table 8.11 Efficiency measure of the pairings

	F1	F2	F4	F6	F8	F9	F10	S1	S3	S4	S5	S6	S7
W1		P		E					E			P	
W2	P												
W4				P		P							
W5			E			E							
W6						E		E			E		
W7								E		E		E	

Table 8.12 Linear programming derived pairings

	F1	F2	F4	F6	F8	F9	F10	S1	S3	S4	S5	S6	S7
W1										2	2		
W2						2							
W4				1		1							
W5								2	1				
W6	2	1	2									1	
W7				1				1					1

The model is:

$$\sum W F_{ij} + \sum W S_{ij} = W D_i \text{ all } i$$

$$\sum W F_{ij} = F D_i \text{ all } j$$

$$\sum W S_{ij} = S D_i \text{ all } j$$

Minimise

$$\sum (R W F_{ij} x W F_{ij} + R W S_{ij} x W S_{ij})$$

The solution produced from this model is shown in Table 8.12.

The efficiency is shown in Table 8.13 at a cost of 2 inefficient pairings more efficient than the plan issued by the company planners (5 inefficient pairings) but still an inefficient solution. This example demonstrating that the production department will always have been given plans where they were not able to produce the standard production quantities.

8.2.6   Implementation of Daily Pair Selection

Following this investigation the company purchased additional machinery and as a result the pairing Table 8.8 could now be replaced with Table 8.14

Table 8.13 Efficiency of LP derived pairings

	F1	F2	F4	F7	F8	F9	F10	S1	S3	S4	S5	S6	S7
W1										E	E		
W2					E								
W4				P		P							
W5								E	E				
W6	E	E	E									E	
W7				E				E					E

Table 8.14 Viability of production pairings with additional resources

	F1	F2	F3	F4	F5	F6	F7	S1	S2	S3	S4	S5	S6	S7	S8
W1	*	*	*	I	*	*	I	*	*	*	*	*	*	*	*
W2	*	*	*	*	*	*	*	*	*	*	*	*	*	*	*
W3	*	*	*	*	*	*	*	*	*	*	*	*	*	*	*
W4	*	*	*	*	*	*	*	I	*	I	I	I	*	I	I
W5	*	*	I	*	I	*	I	*	*	*	I	*	*	*	*
W6	*	*	I	*	I	*	I	*	*	*	I	*	*	*	*
W7	*	*	*	I	*	*	I	*	*	*	*	*	*	*	*
W8	*	*	I	*	I	*	I	*	*	*	I	*	*	*	*
W9	*	*	I	*	I	*	I	*	*	*	I	*	*	*	*
S1	*	*	*	*	P	*	*								
S2	*	*	*	*	*	*	*								
S3	*	*	*	*	*	*	*								
S4	*	*	*	*	P	*	*								
S5	*	*	*	*	*	*	*								
S6	*	*	*	*	*	*	*								
S7	*	*	*	*	P	*	*								
S8	*	*	*	*	*	*	*								

Leading to:

- Possible Pairs 191
- Efficient Pairs 162, (85%)
- Inefficient Pairs 3, (1%)
- Impossible Pairs 26, (14%)

Table 8.15 Efficiency of production with additional resources

	F1	F2	F4	F6	F8	F9	F10	S1	S3	S4	S5	S6	S7
W1		E		E					E			E	
W2	E												
W4				E		E							
W5			E			E							
W6						E		E			E		
W7								E		E		E	

and considering the reduced table, only WF and WS combinations:

- Possible Pairs 135
- Efficient Pairs 109, (81%)
- Inefficient Pairs 0, (0%)
- Impossible Pairs 26, (19%)

Consequently, the previously inefficient set of production pairs are now rated as being efficient and the production unit will have been able to demonstrate efficient operation; see Table 8.8, and this showing that the pairing used in Table 8.15 now enables the production unit to achieve its production target.

As result of the purchase of additional production resources the average weekly production rose by 10% and the wastage rate fell from 18% to 2%.

More importantly the quantity of usable products manufactured in this unit rose by 30%, this change followed from:

- The managers insistence on the production of quality items
- Fully resourced department
- Better planning

Because the department capacity, with the current staffing, is much in excess of the current demand the company could:

- Increase marketing of these products to boost demand and use this excess capacity to satisfy the increased demand, or
- Plan production in half days, less standard clean up times, enabling a more flexible response to changes in demand. Provided that the planning office makes an appropriate adjustment to the expected daily production to allow for this reduction in available time.

8.3   Impact of Lack of Information on Company Profits

A measurable effect on the company's profits resulting from their implementation of inefficient production plans was the reduction in contribution to profit from the sales of these (HB) items.

This can be seen through an analysis of their pricing structure.

The price and standard production quantities from each of their products are given in Table 8.16.

The daily production quantities having been determined assuming that all production was efficient, however as has been shown production was inefficient not being able to meet production targets (resulting from data problems); consequently, costs were higher than expected and contribution to profit reduced.

The company expected that the return from the sales of each line would have the form:

$$R_i = (a + bX_i) + P$$

where

- R_i is the profit from a standard production day
- Y_i is the standard day production quantity

Table 8.16   Costing data

Product types	Number of lines	Unit price	Daily production
A	1	19	2.110
B	1	19	1.984
C	3	19	2.140
D	2	19	2.380
E	1	19	1.920
F	1	19	2.280
G	1	24	1.360
H	2	21	1.880
J	2	17	3.000
K	1	17	2.520
L	1	19	2.250
M	1	19	2.460
N	1	19	1.920
P	2	17	2.590
Q	1	17	2.430
R	1	19	2.450
S	1	19	1.330

- $X_i = \frac{1}{Y_i}$ is the time required to produce a standard production unit
- P is the required profit from a standard day production quantity
- a and b are constants

Using this data, omitting item S which was obviously mis-priced, it follows that price (P) and daily production (D) are related through:

$$R = 11 + 17.4X$$

and therefore because the actual production quantities, for many lines, are, and had to be, less than the standard production quantities the quoted prices are too low and the units contribution to profit will be lower than planned.

For example if for line A the achievable daily quantity had been 1.90 rather than the expected 2.11 then the unit cost would have been 20 rather than 19, leading to a loss of 5% in the contribution to profit.

A result of the lack of accurate data about production and a lack of data flow through the firm.

8.4 Conclusion

The real problems within this firm were all concerned with, or influenced by, the lack of timely information flowing through and between the various departments resulting in:

- The firm not knowing the 'real' sales demands and 'real' stock levels when producing a production plan;
- The firm not knowing the actual daily production capacities meant that in many cases the daily production targets were too optimistic;
- The firm not knowing the actual daily productions;
- Also working with the available aggregated sales figures meant that the retail outlets either received too much new stock or too little new stock;

These, and other, problems could have been alleviated through use of available computer-based data enabling the planners to produce meaningful production schedules and the firm being aware of potential problems, errors in standard daily production quantities.

8.5 Learning Activities

? Exercise 1

In this exercise we shall be comparing manufacturing systems.

N&F plc operates a small workshop in Ayton producing specialist items. The workshop currently contains 6 machines and has had a workforce of 6, one worker for each machine. Currently jobs arrive randomly with, on average, an arrival every 12 min.

The firm is about to reorganise their production facilities either by buying new machines or by establishing production lines using their existing machines. The current system uses the machines organised as the Flow Shop as shown in Fig. 8.6.

Alternative proposals:

Proposal 1: Replace the machines at stages B and C with two multipurpose machines (M), job time between 10 and 15, replacing two production stages with one stage (Fig. 8.7).

Proposal 2: Set up two production lines each (assume that there are now two machines of type C and 2 machines of type A), buy an extra machine type C as in Fig. 8.8.

With one machine at each stage and jobs allocated at random to each of these production lines.

Proposal 3: Set up two production units with a common final stage where 66% of the jobs are allocated to production unit X (Fig. 8.9).

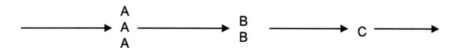

Fig. 8.6 Six-machine manufacturing system

Fig. 8.7 Alternate arrangement of manufacturing system (Proposal 1)

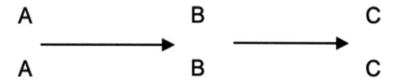

Fig. 8.8 Alternate arrangement of manufacturing system (Proposal 2)

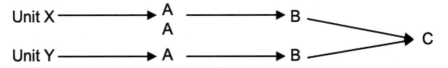

Fig. 8.9 Alternate arrangement of manufacturing system (Proposal 3)

Table 8.17 Production data

Machine type	Job time distributions average times in minutes exponential or deterministic	Job times uniform distribution times in minutes
A	15	Between 10 and 20
B	6	Between 5 and 7
C	5	Between 3 and 6
M	9	Between 8 and 10

Using the production data in Table 8.17 construct queueing simulation models to evaluate these alternative systems.

? Exercise 2

Production planning at a fireplace manufacturing company
Evans Fireplace Factory produces a range of fireplace surrounds from engineered composite wood (Medium Density Fibreboard, MDF), hardwood and stone at its factory. The hardwood surrounds usually incorporate some individual design features and are (thus) mainly produced to satisfy (individual) customer sales while some MDF surrounds, which is a more standard product, can be held in storage against future demand for this product.

Evans is also developing a partnership with a regional chain of DIY outlets to supply MDF fireplaces kits for sale through their outlets and as a consequence of this partnership agreement with W&B the loading of the production facilities at Evans are approaching capacity resulting in extended, unsatisfactory, delivery lead times and a (slight) decline in product quality (increasing customer complaints).

The production process used varies with the type of product made; however the equipment used is given in Tables 8.18 and 8.19 with the layout of the production system as shown in Fig. 8.10.

Table 8.18 Current manufacturing process data

| | Time (min) | | |
	Softwood	Hardwood	Stone
Prepare wood	30 ± 10	120±15	240±60
Make decorations	10 ± 2	15±5	Not Relevant
Construct frame	20±10	40±15	15
Finish frame	15±1	30±1	Not Relevant
Add decorations	10±1	10±1	Not Relevant

Table 8.19 Proposed manufacturing process

| | DIY chain product time (min) | | |
	Softwood	Hardwood	Stone
Wood&decorations	45±15	140±10	240±60
Construct frame	20±15	40±15	15±5
Finish frame	Off site	–	–

a) Current production flow

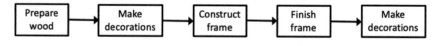

b) Proposed production flow

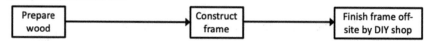

Fig. 8.10 Layouts for **a** current processing arrangement and **b** proposed alternative

8.5.1   Machinery and Staffing

Currently there are four production areas (either machines or work tables):

- Preparation—3 work tables;
- Decorations—1 machine;
- Construction—2 work tables;
- Decorations and finishing—3 work areas;
- and 5 production workers.

Table 8.20 Probability of a breakdown and the repair times

| | Breakdown probability and repair times (min) | | |
	Softwood	Hardwood	Stone
Prepare	0.075; 5±1	0.05; 10±3	Not relevant
Make frame	0.1; 8±1	0.1; 8±2	Not relevant

8.5.2 Sales Data

Own Sales: Currently, during the peak demand seasons, each day the manufacture of 10 units is planned, normally 6 softwood, 3 hardwood, and 1 stone surround using this current factory layout. The company hopes that sales will increase.

Partnership Sales: Evans partnership with W&B has resulted in a demand for on average 25 MDF kits a week, the production and installation of these surrounds are described by table and figure C3.2 where the finishing tasks (adding decorations and final painting) are carried out by the DIY chain's workers.

Note if this new product is successful the DIY chain intends to make hardwood and stone surrounds available through their outlets.

Part 1: Proposed system for the DIY kits
Assuming that all workers are multi-skilled construct a simulation model to represent the proposed production system as described in Fig. 8.10 and use this model to determine an appropriate number of machines at each stage and the staffing required to be able to satisfy both own and DIY sales.

Part 2
For each of the models developed in Part 1, investigate the effect of machine breakdowns and repairs on your recommended model.

The breakdown and repair information are given in Table 8.20.

? Exercise 3

Evaluation of alternative service systems at a retail bank
Swifts Bank operates its retail banking services through a chain of high street outlets. To be able to improve customers service, and reduce costs, Swifts are considering the reorganisation of their branch counter services.

Customers arrive randomly, on average one every 1.5 min. 60% of these customers are classified as 'short service time customers', 30% are 'medium service time customers', 9% are 'long service time customers' and the remaining 1% require in-depth advice.

The service times for these customers are:

- *Short service*—between 1 and 3 min
- *Medium service*—between 2 and 6 min
- *Long service*—between 5 and 15 min
- *In-depth advice*—between 10 and 30 min

Construct appropriate queueing and/or simulation models to evaluate the performance of current service system, considering customer waiting time, queue length,
server idle time and number of servers needed to provide the customers with a good
service and hence determine the most appropriate system for use by Swifts Bank.

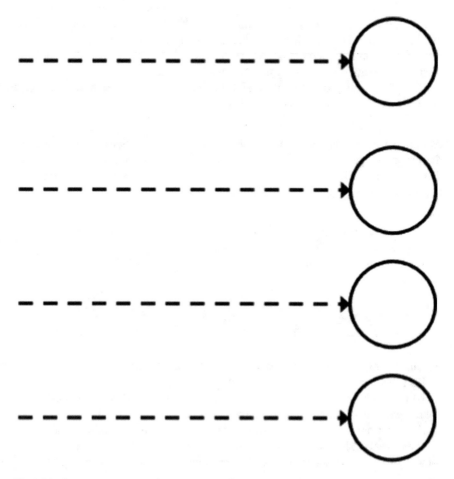

Fig. 8.11 Current arrangement: four queues and four servers

Notice that an essential task is the definition of:

- What constitutes good service, customer time in system
- What constitutes efficient service (bank), costs

Part 1: Evaluating the current system
Currently there are four service counters in use with queues of waiting customers forming in front of each server; see Fig. 8.11; you may assume that there is no queue switching.

Part 2: Evaluating first alternative system
A first possible reorganisation is to establish two queues one serving a new quick service counter (only quick jobs) and the other serving the other three counters; see Fig. 8.12.

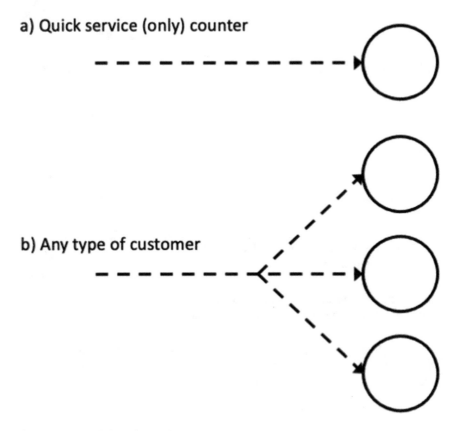

Fig. 8.12 Queues in bank for service

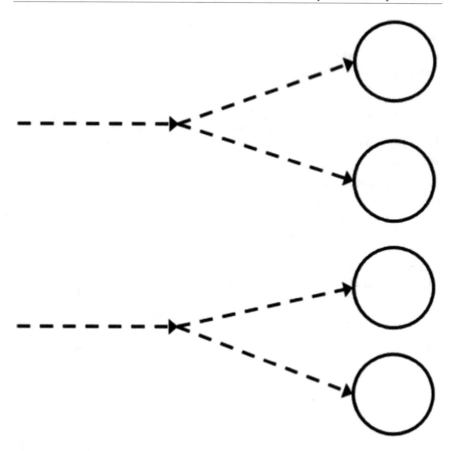

Fig. 8.13 Two queue service

Part 3: Evaluating second alternative systems
The possible reorganisations are:

- to establish two queues each serving two counters,
- to establish one queue serving all counters; see Figs. 8.13 and 8.14a.

Part 4: Implement a reception (triage) counter to direct customers
A receptionist directs the customers to the queue in front of the most appropriate server; see Fig. 1d, duration 90% 1 min, 10% 2 min.

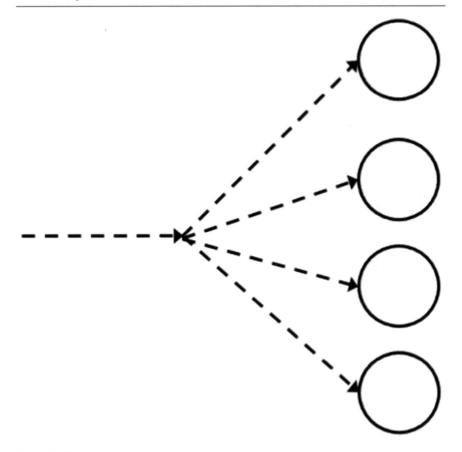

Fig. 8.14 One queue, four servers

Minimum Information Set for Effective Control

9

9.1 Information Flows Within an Organisation

Within an organisation information should flow between the different functions (sales, manufacturing, inventory control, dispatch) and the volume and quality of this information will affect the ability of the firm to operate successfully (profitably) making and efficient and effective use of the production facilities to be able to supply (on time) good of the required quality.

However problems in production could arise, resulting from a lack of stock of subcomponents or raw materials, when, as in Fig. 9.1, sales does not directly involve information from production planning or inventory control when setting a delivery time/job completion time. Here there will be very poor data available to sales to enable the agreement of realistic delivery dates and the consequences of the unavailability of good data would be that:

- the firm will not be able to set appropriate (optimal) delivery dates and would (often) quote a standard fixed (long/safe) delivery time to ensure that the finished good will be available on time for the customer, and;
- there can be high inventory costs as the stock controller keeping excessive stock just in case demand is higher than normal.

The planning process can be improved, leading to an *Order Book* planning system, when production feeds back information to sales concerning the status of jobs within the system (job completion times/time in production, see Fig. 9.2) thus enabling the firm to set an optimal (or close to optimal) delivery date allowing an efficient use of resources and supplying customers on time.

More and better data about the status of the system, to enable better planning, will be generated when the information flow model shown in Fig. 9.2 are enhanced as in the amended model shown in Fig. 9.3 where there are more direct connections between the various departments.

© Springer Nature Switzerland AG 2021

R. Hill and S. Berry, *Guide to Industrial Analytics*, Texts in Computer Science,

https://doi.org/10.1007/978-3-030-79104-9_9

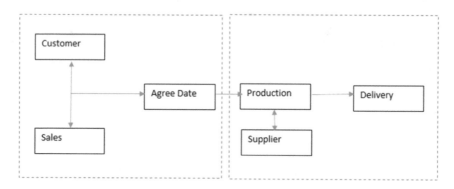

Fig. 9.1 Disjointed system

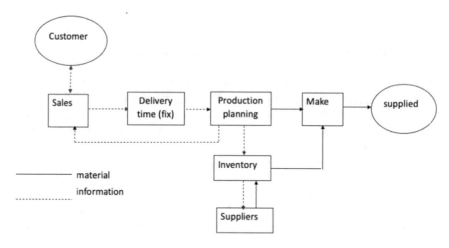

Fig. 9.2 Model of information

   The data flow diagram, Fig. 9.4, indicates the data gathering that may be required to enable a reliable forecast for the expected job completion time for the newly agreed job.

   Similarly data sharing also enables the collation of the information necessary to enable an effective inventory control system. Thus reducing inventory costs while still enabling the efficient and effective use of the production facilities to supply the customer on time, Fig. 9.5.

   Data sharing leads enables the establishment of an effective, and feasible, production plan and to an efficient (low cost) inventory policy. The use of this shared data in these planning processes is illustrated in Fig. 9.6 and the quality and quantity of data used illustrated in Fig. 9.7.

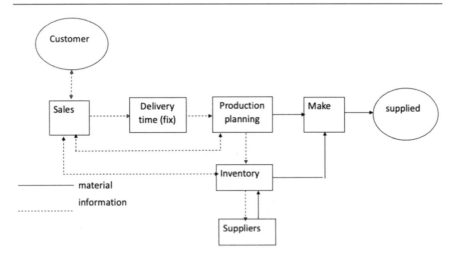

Fig. 9.3 Improved model of information

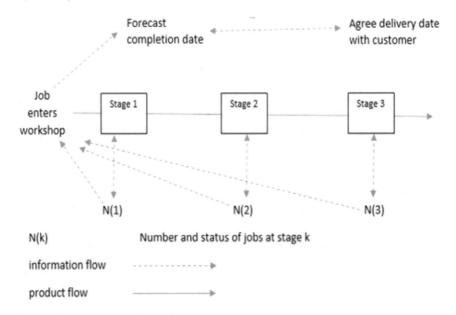

Fig. 9.4 Data collection and analysis

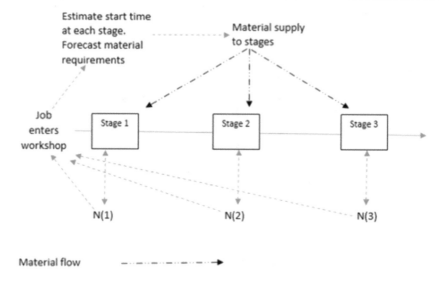

Fig. 9.5 Planning materials requirements

9.2 Deriving Minimum Information Requirements

The quantity and quality of the available information about the status of a production system, at the time of arrival of a new job, would be expected to influence the quality of the forecast for its completion time, to be quoted by sales to the customer.

The objective, here, is to investigate, through the construction of simulation models, the required quality and quantity of data needed to be able to reliably estimate job completion times within a Flow Shop environment. The initial investigation is concerned with the evaluation of 'order book'-based approaches to the estimation of job completion times and the second part is concerned with the evaluation of a 'work book'-based approach.

In each case deriving the minimum information set required to be able to produce good estimates (in all cases investigated here it is assumed that the firm adopts a greedy approach to the allocation of production time, see Appendix D for a consideration of a workshop employing a KANBAN-based allocation of production time).

9.3 Order Book-Based Systems

The investigation into order book-based systems commences by considering two workshop configurations, each containing three processors in sequence described by the notation $C_3[1, 1, 1]$ Fig. 9.8, there are three stages with one processor at each stage and within each sequence a machine is described as being dominant (D) if its

Fig. 9.6 Processing and sharing data

average processing time is greater than the average times at any other stage. Stage k is dominant if, the average time at stage k is such that:

$$T_k > T_j \text{ for } j < k, \text{ and } T_k \geq T_j \text{ for } j \geq k$$

The configurations investigated are:

- dominant machine first $\{C_3[1, 1, 1], \text{DNN}\}$ and;
- dominant machine last $\{C_3[1, 1, 1], \text{NND}\}$

and for each configuration investigating those workshops where variability to the product is added at:

- each stage within the production process; or
- the final stage (only) in the production process all earlier stages common to all products.

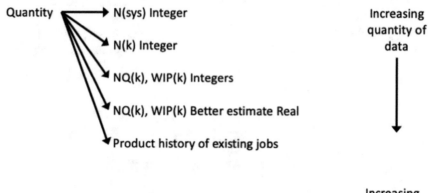

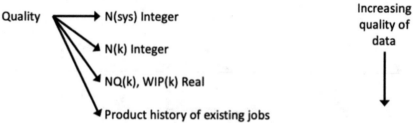

Fig. 9.7 Data quality and quantity used in forecasting.

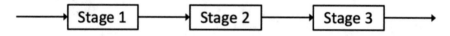

Fig. 9.8 Base model of three production stages

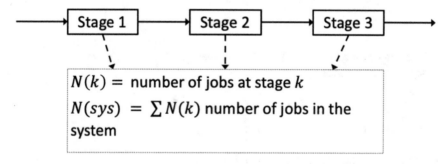

Fig. 9.9 Data gathering

At the arrival of a new customer at sales the following information could be made available (using the information available from an order book) to determine a possible delivery date for this new job, Fig. 9.9.

Notice that the number of jobs at a stage comprises the number queueing and the number being served where:

- $NQ(k)$—the number queueing at stage k will be an integer;
- $NP(k)$—the number being processed at stage k could be integer valued $\{0,1\}$ or real valued 0 to 1 if its production completion status is known.

Thus, for an order book (OB)-based methodology to derive a forecasting model it follows that:

- minimal information set—$N(sys)$
- intermediate information set—$NQ(k), NP(k)$ as integer
- maximum information set—$NQ(k)$ integer and $NP(k)$ real valued.

In each case either the average processing time at each stage, or the distributions of times at each stage are known.

9.3.1   Deriving an Order Book (OB) Forecasting Model Where Orders for New Jobs Arrive Randomly in Time

For the two workshop configurations two variants were considered:

- workshops where variability to the product is added at each stage and;
- workshops where variability to the product is added at (only) the final stage.

Three hypotheses were tested:

- H_1—the derived models are the same for both configurations;
- H_2—the derived models are independent of the demand levels;
- H_3—the models can be derived using only the average stage processing time.

The base-level case consisting of three processors in series with parameters given in Table 9.1 and workshops with configurations:

- ABC configuration DNN dominant, longest, stage first, $\{C_3[1, 1, 1], DNN\}$
- BCA configuration NND dominant stage last, $\{C_3[1, 1, 1], NND\}$

A rectangular (uniform) distribution was chosen to represent a firm having a minimum of good information available about the behaviour of the processes at each stage, and these distributions have a high standard deviation to emphasise forecast errors. A job processed without any delay (queueing between stages) could have a duration between 17 and 25, average process time $\pm19\%$.

Table 9.1 Service time distributions

Processor	Job time distribution	Maximum	Minimum	Mean	Standard deviation
A	Uniform	9	7	8	0.58
B	Uniform	8	5	6.5	0.87
C	Uniform	8	5	6.5	0.87
A, B and C in series				21	1.36

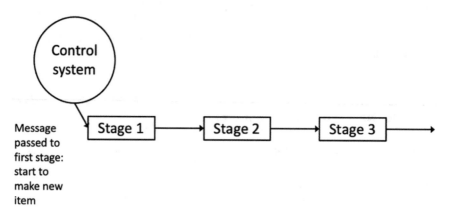

Fig. 9.10 Control passes message to start production

Note Defining Dominance:

- strong dominance, stage k is strongly dominant if Minimum(t_k) > Maximum(t_j) all $j \neq k$
- weak dominance, stage k is weakly dominant

$\quad$ If $\bar{t}_k \geq \bar{t}_j$ all $j \neq k$ but Minimum t_k < Maximum t_j for $j = i$

9.3.2   Variability in the Final Product Introduced at All Stages

Here at the receipt of an order for a new job a message will be passed onto the first stage within the production process to prepare to initiate this new job, see Fig. 9.10, and assuming that the number of jobs present $N(sys)$ is known, loadings of 80 and 66% (representing high and medium levels of demand) were used to investigate Hypotheses 1 to 3.

9.3.3 Simple Order Book System

A simple system will assume that there is a linear relationship between the number of jobs in the system and the expected completion
time for a newly arrived job. Here two variants are investigated:

- simplest data using N(sys), total number of jobs in the system
- enhanced data using N(1), number of jobs at the first stage

Case 1: NND using N(sys) 66% loading at the dominant stage The model produced from the simulation of this system, using only average job times, is:

$$JCT = 13.40 + 2.20\,N(sys)$$

For comparison the model produced where job times are randomly drawn from a distribution is:

$$JCT = 14.46 + 5.68\,N(sys)$$

Case 2: DNN using N(sys) 66% loading at the dominant stage.
The model produced from the simulation of this system, using only average job times, is:

$$JCT = 14.75 + 6.20\,N(sys)$$

For comparison the model produced where job times are randomly drawn from a distribution is:

$$JCT = 14.49 + 5.81\,N(sys)$$

Notice that the accuracy of the data used does impact upon the models produced but using 'best data' within the simulations leads to a consistent model.

However the forecast errors (comparing forecast with actual) range between -40 and 50%, see Fig. 9.11, indicating that this linear model does not provide a reliable approach to forecasting job completion times.

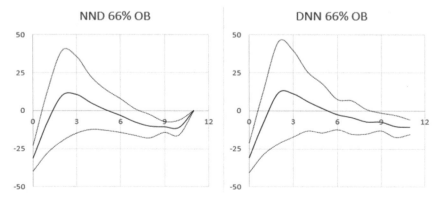

Fig. 9.11 Plotting percentage errors

Case 3: DNN and NND using N(1) 80% loading at the dominant stage
The models produced for the DNN configuration were:

$$\text{Job Time} = 18.34 + 7.81(\text{Work in System})$$

However when there is 'reasonable' knowledge concerning the form of the service distributions then 'fixing' as a base value the average service time when there are no job in the system gave the model:

$$\text{Job Time} = 21.00 + 7.66(\text{Work in System})$$

No reasonable models were produced for the NND configuration at this loading:

- H_1—False, the models produced differ significantly;
- H_2—False, demand does not significantly affect the derived model;
- H_3—False, the models produced differ significantly.

9.3.4 Enhanced Order Book Models

An alternative approach would be to acknowledge that a two-phase model would better represent the data as per Fig. 9.12.

To test H_1, H_2 and H_3 against this model, loadings of 66% (medium demand) and 50% (low demand) were simulated with both configurations, $\{C_3[1, 1, 1]\text{DNN}\}$ and $\{C_3[1, 1, 1]\text{NND}\}$.

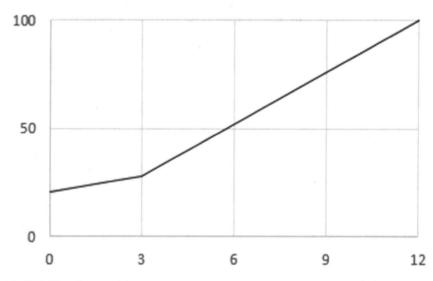

Fig. 9.12 Two-phase model

Case 1: NND using the Minimal Information set, demand loading 66%

The model produced from the simulation of this system, using only average job times, is:

$$N(sys) \leq 3JCT = 21 + 2.33N(sys)$$

$$N(sys) \geq 3JCT = 4 + 8.01N(sys)$$

For comparison the model produced where job times are randomly drawn from a distribution was:

$$N(sys) \leq 3JCT = 21 + 2.42N(sys)$$

$$N(sys) \geq 3JCT = 4 + 8.28N(sys)$$

Case 2: DNN using the Minimal Information set, demand loading 66%.

The model produced from the simulation of this system, using only average job times, is:

$$N(sys) \leq 3JCT = 21 + 2.33N(sys)$$

$$N(sys) \geq 3JCT = 4 + 8.06N(sys)$$

For comparison the model produced where job times are randomly drawn from a distribution was:

$$N(sys) \leq 3JCT = 21 + 2.34N(sys)$$

$$N(sys) \geq 3JCT = 4 + 8.24N(sys)$$

Case 3: NND using the Minimal Information set, demand loading 50%.

The model produced from the simulation of this system, using only average job times, is:

$$N(sys) \leq 3JCT = 21 + 2.40N(sys)$$

$$N(sys) \geq 3JCT = 4 + 8.29N(sys)$$

Case 4: DNN using the Minimal Information set, demand loading 50%.

The model produced from the simulation of this system, using only average job times, is:

$$N(sys) \leq 3JCT = 21 + 2.22N(sys)$$

$$N(sys) \geq 3JCT = 4 + 8.17N(sys)$$

These results indicating that:

- H_1—TRUE: the derived models are independent of system configuration;
- H_2—TRUE: the derived models are independent of system demand;
- H_3—TRUE: the models can be derived using deterministic (average processing) time.

9.3.5   Forecast Accuracy/Validation

Repeating the simulations and comparing the new actual values with the forecasts obtained using these derived models gave the percentage error plots (66% loading in each case) (Fig. 9.13) showing a plot of the average percentage error and the two standard deviation error bounds for all values of N(sys) (number of jobs present).

The state of maximum uncertainty occurring when:

The number of Jobs $\approx$ Present Number of Production Stages

Thus suggesting that (here) that when 2, 3 or 4 jobs are present more information would be needed to enable the production of a better estimate (reducing percentage error).

A further investigation considered the use of less data, number of jobs at the first stage (only) leading to the hypothesis:

H_4 — using Stage 1 only will improve forecast

The results showed that for the configuration $\{C_3[1, 1, 1], \text{DNN}\}$ using N(1) job completion time can be modelled by:

$$N(sys) \leq 1 JCT = 21 + 6.61N(1)$$

$$N(sys) \geq 1 JCT = 4 + 7.76N(1)$$

These indicate that the state of maximum uncertainty occurs when

$$N(sys) \approx \text{Number of Production Stages}$$

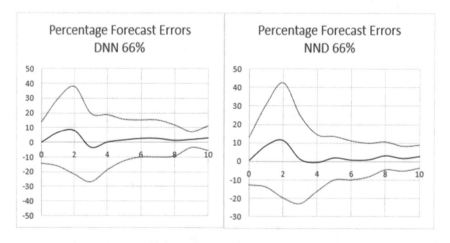

Fig. 9.13 Plots of the average percentage error and the two standard deviation error bounds for all values of N(sys) (number of jobs present)

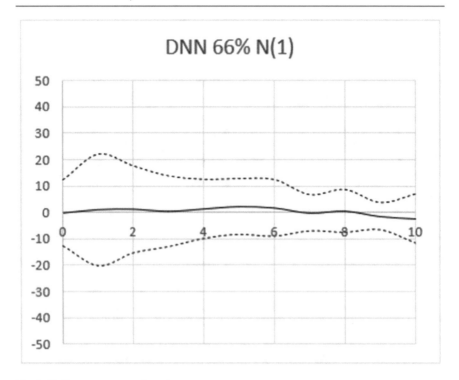

Fig. 9.14 Percentage error plot

Percentage error plots indicating that the errors using N(1) or N(sys) were similar, H_4 False (Fig. 9.14).

However, for the configuration $\{C_3[1, 1, 1], NND\}$ there was no appropriate model using only data from the first stage.

Although consistent models have been generated such order book-based models tend to produce unreliable forecasts (high percentage errors) when there are few jobs within the system suggesting that a larger database would be required to enable reliable forecast.

9.3.6   Extending the Investigation by Including Data from All Stages

The extended (OB) model includes data from all stages to construct multiple linear regression-based models. Thus, leading to the additional hypotheses:

- H_5—using data from all stages will improve the DNN forecast
- H_6—using data from all stages will improve the NND forecast

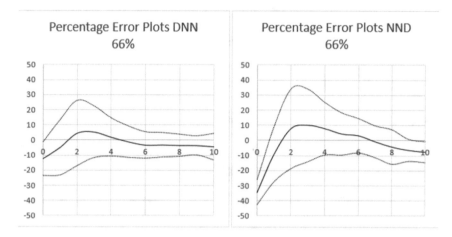

Fig. 9.15 Percentage error plots using data from all stages

Multiple linear regression models were constructed for each configuration:

$$C_3[1, 1, 1], \text{ DNN } JCT = 18.25 + 7.22N(1) + 1.62N(2) + 0.42N(3)$$

$$C_3[1, 1, 1], \text{ NND } JCT = 14.93 + 7.37N(1) + 4.52N(2) + 5.23N(3)$$

These models demonstrated differences between the forecasting (data) needs for these two (basic) configurations.

For the DNN configuration, the number of jobs at Stage 1 is dominant while for the NND configuration all stages need to be considered.

Thus the optimal (OB) approach will be dependent upon the location of the dominant stage within the production system.

Notice the reduction in the maximum percentage errors (Fig. 9.15), compared with the results from the earlier models, but the worst-case errors are still large.

The results indicating that both H_5 and H_6 are False.

For comparison, if 'best data' were available, estimate of remaining work at each stage graphs (Fig. 9.15) is replaced with the graphs in Fig. 9.16.

9.3.7   Variability Added (only) at the Final Stage

When a firm makes a single product, with many variants, and although the production times at a stage all follow the same distribution each variant will have a slightly different manufacture (different components) at only some of the production stages, the initial stages (often) being common with all items as a consequence it will be feasible to carry some work in progress at these common stages.

Here variant discrimination occurs (only) at the final stage within the production process and as a consequence 'start production' messages will be sent to the first and final stages in the production process (Fig. 9.17).

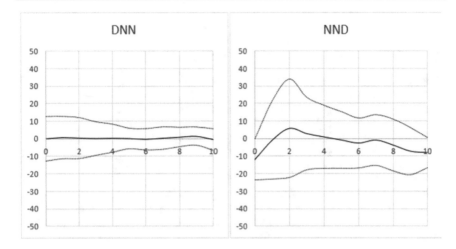

Fig. 9.16   Percentage error plots using data from all stages, using 'best data'

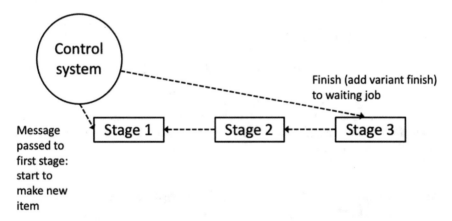

Fig. 9.17   Messages passed to first and variant stages

Simulating the performance of this system, for both $C_3[1, 1, 1]$, DNN and $C_3[1, 1, 1]$, NND configurations, at a demand level equivalent to a loading of 67% allowed work in progress (at stages 1 and 2) leads to the models:

Configuration $C_3[1, 1, 1]$, DNN

$$N(3) \leq 1 JCT = 6.48 + 3.62N(3)$$
$$N(3) \geq 1 JCT = 4.08 + 5.97N(3)$$

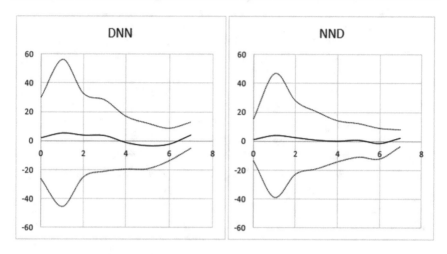

Fig. 9.18 Percentage error plots

Configuration $C_3[1, 1, 1]$, NND

$$N(3) \leq 1 \, JCT = 8.00 + 4.56N(3)$$
$$N(3) \geq 1 \, JCT = 4.80 + 7.62N(3)$$

The error plots indicated (Fig. 9.18) that more and/or better information would be required to be able to make a better forecast on those occasions when there are fewer jobs within the system.

9.3.8 Conclusion: Order Book (OB)-Based Approaches to Forecasting

The results from these investigations demonstrate that when variability is added to the product at all stages within the production process the location of the dominant stage within the production process is an important factor influencing the quantity and quality of data required to implement an efficient and effective production planning and control system.

In general, a system with the configuration DNN (dominant first) requires little (approximate) data to derive a forecasting model while a system with the configuration NND requires more and better quality data to be able to derive such a model. These results also show that this approach (OB) does not produce as reliable result (greater forecast errors) when variability is added to the product at only the final stage of the production process.

9.4   Work Book (WB) Systems

The data requirement to be able to implement this approach is that at the arrival of a
new job there exists records of the progress of all previous jobs through the system:

- the number of jobs active at each stage (queueing or being processed) and;
- the start and finish times for each of these jobs at each stage when known.

For example given that at the arrival of a new job the state of the system is as
shown in Fig. 9.19.

A total of five jobs is currently active within the production system at the arrival
of this new job. The known status of each of these jobs at each stage in the production
system is shown in Table 9.2).

Table 9.3 is produced through the addition of the expected stage times where the
actual times are not yet known (job not yet completed at the stage although job
starting time may be known).

Table 9.4 shows the results from the simulation of the system and the expected
completion time for this new job (and for all the jobs currently present within the
system).

Table 9.5 shows the 'real' completion times, for comparison with Table 9.4.

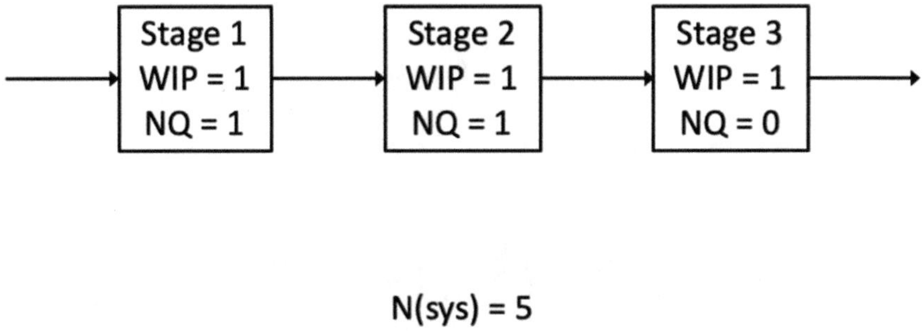

$$N(sys) = 5$$

Fig. 9.19   Base model

Table 9.2   Status of current jobs

Job number	Arrival time	Position	Stage 1	Stage 2	Stage 3
0	Known	Known	Known	Known	Known
1	Known	WIP(3)	Known	Known	–
2	Known	WIP(2)	Known	–	–
3	Known	Q(2)	Known	–	–
4	Known	WIP(1)	–	–	–
5	Known	Q(1)	–	–	–
New	TNow	Joining	–	–	–

Table 9.3 Available and known stage data for all current jobs

S1				S2				S3				
Arrival	Duration	Start	End	Arrival	Duration	Start	End	Arrival	Duration	Start	End	Duration
2127.9	6.3	2136.1	2142.3	2142.3	7.6	2143.8	2151.4	2151.4	7.3	2154.2	2161.5	33.6
2129.7	6.9	2142.3	2149.2	2149.2	5.3	2151.4	2156.6	2156.6	8.0	2161.5		
2140.5	6.7	2149.2	2155.9	2155.9	6.5	2156.6			8.0			
2141.1	6.4	2155.9	2162.2	2162.2	6.5				8.0			
2150.2	6.5	2162.2			6.5				8.0			
2154.3	6.5				6.5				8.0			
2162.9	6.5				6.5							

Table 9.4 Calculating estimated completion times

S1				S2				S3				
Arrival	Duration	Start	End	Arrival	Duration	Start	End	Arrival	Duration	Start	End	Duration
2127.9	6.3	2136.1	2142.3	2142.3	7.6	2143.8	2151.4	2151.4	7.3	2154.2	2161.5	33.6
2129.7	6.9	2142.3	2149.2	2149.2	5.3	2151.4	2156.6	2156.6	8.0	2161.5	2169.5	39.8
2140.5	6.7	2149.2	2155.9	2155.9	6.5	2156.6	2163.1	2163.1	8.0	2169.5	2177.5	37.0
2141.1	6.4	2155.9	2162.2	2162.2	6.5	2163.1	2169.6	2169.6	8.0	2177.5	2185.5	44.4
2150.2	6.5	2162.2	2168.7	2168.7	6.5	2169.6	2176.1	2176.1	8.0	2185.5	2193.5	43.3
2154.3	6.5	2168.7	2175.2	2175.2	6.5	2176.1	2182.6	2182.6	8.0	2193.5	2201.5	47.2
2162.9	6.5	2175.2	2181.7	2181.7	6.5	2182.6	2189.1	2189.1	8.0	2201.5	2209.5	46.6

Table 9.5 Actual completion times for comparison with estimated times

S1						S2						S3					
Arrival	Duration	Start	End			Arrival	Duration	Start	End			Arrival	Duration	Start	End		Duration
2141.1	6.4	2155.9	2162.2			2162.2	7.4	2163.2	2170.6			2170.6	8.3	2176.2	2184.5		43.4
2150.2	7.0	2162.2	2169.2			2169.2	7.2	2170.6	2177.7			2177.7	7.9	2184.5	2192.4		42.2
2154.3	6.1	2169.2	2175.1			2175.4	6.0	2177.7	2183.7			2183.7	7.3	2192.4	2199.6		45.3
2162.9	5.8	2175.4	2181.2			2181.2	7.4	2183.7	2191.2			2191.2	7.7	2199.6	2207.3		44.5

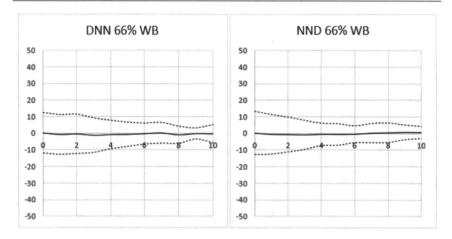

Fig. 9.20 Work book using data from all stages

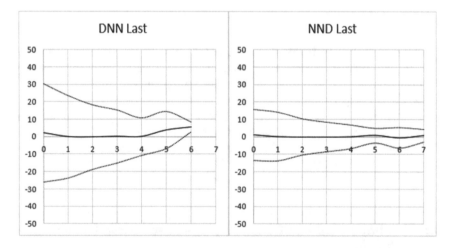

Fig. 9.21 Work book using data from all stages (variability added at final stage)

This simulation indicating that the forecast finishing time for this new job is at time 2209.5 with an expected duration, time in system, of 46.6 compared with an actual finishing time of 2207.5 and an actual duration of 44.5 (percentage error of 4.4%).

The simulation also provides updated estimates for the job completion times for all jobs currently within the system, allowing effective control.

Applying this approach when variability is added to the product at all stages within the production system gave the error plots shown in Fig. 9.20.

Applying this approach when variability is added to the product at only the final stage within the production system gave the error plots shown in Fig. 9.21.

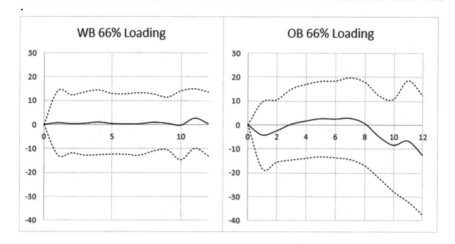

Fig. 9.22 Comparison of error plots for work book (WB) and order book (OB) approaches to planning

These results demonstrating that a WB-based approach can give very good forecasts for job completion times except in the case when variability is added at only the final stage and the workshop has the configuration DNN.

As a final comparison the error plots for a larger workshop $\{C_3[1, 1, 1]DNN\}$ with variability added at each stage) using both the Work Book and the order book approach to forecasting job completion times are shown in Fig. 9.22. These plots again confirm that the best approach would be to use a WB system thus giving the implied data requirement.

9.5 Evaluating WB and OB When Stage Productions Have Been Balanced

In some workshops the number of processors at each stage has been chosen so that the capacities at each stage are (approximately) equal, leading to a system where there is no (very) dominant stage.

For example, a three-stage process is illustrated in Table 9.6.

Table 9.6 Processors and average times for each production stage

Stage	1	2	3
Number of processors	5	3	2
Average time at stage	11	7	5

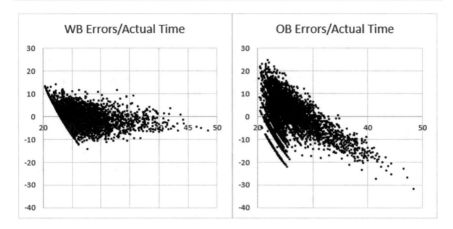

Fig. 9.23 Balanced workshop plotting percentage error against duration

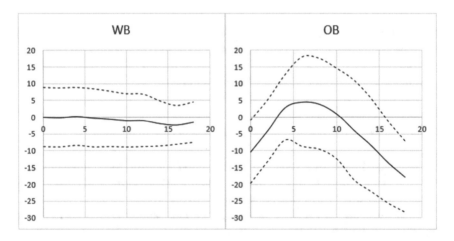

Fig. 9.24 Number in system and two standard deviation bounds on forecast error

Applying both systems when the loading on the system is 62.5% lead to the results summarised in Figs. 9.23 and 9.24.

The graphs indicate that a WB-based system is most appropriate for use in a balanced production workshop.

9.5.1   Conclusions and Recommendations MIR

To compare the two (best) approaches to forecasting job completion time, multiple linear regression using enhanced order book data (number of jobs at each stage within the production process) and the work book approach, using job history data, cumulative probability plots by percentage error in the forecasts are shown in Fig. 9.25, for

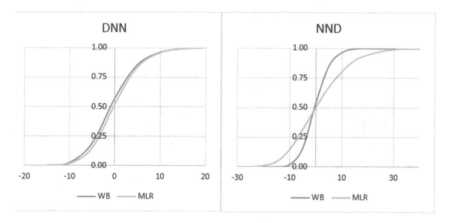

Fig. 9.25 Comparison plots for $\{C_3[1, 1, 1], \text{DNN}\}$ and $\{C_3[1, 1, 1], \text{NND}\}$ at 67% loading

a small firm, and graphs 11.12, for a larger firm where stage capacities are equal. In each case, variability in the final product is added at all stages within the production process.

With respect to the small firm the operation of both:

- $\{C_3[1, 1, 1], \text{DNN}\}$ and;
- $\{C_3[1, 1, 1], \text{NND}\}$ configurations were simulated (at 67% loading).

These plots indicating that the best approach to forecasting job completion times will be dependent upon the configuration of the workshop (location of the dominant stage) and the stages when variability is added to the product

When variability is added at each stage the most appropriate approaches and implied data requirements are:

- $\{C_3[1, 1, 1], \text{DNN}\}$—there is little difference in performance between the two approaches thus the quantity of available data would influence the choice of approach in a given application. The minimum dataset would be the values for $N(k)$, all k and the use of a multiple regression model (if this model has been derived) otherwise the record of job start and finish times at each stage within the production process and a WB approach.
- $\{C_3[1, 1, 1], \text{NND}\}$—the WB system tends to produce more reliable forecasts than an OB approach. The minimum data requirement would be the records for previous jobs and a WB approach. The minimum dataset would be the record of job start and finish times at each stage within the production process.

When variability is added at the final stages the most appropriate approach and implied data requirement is independent of the location of the dominant machine

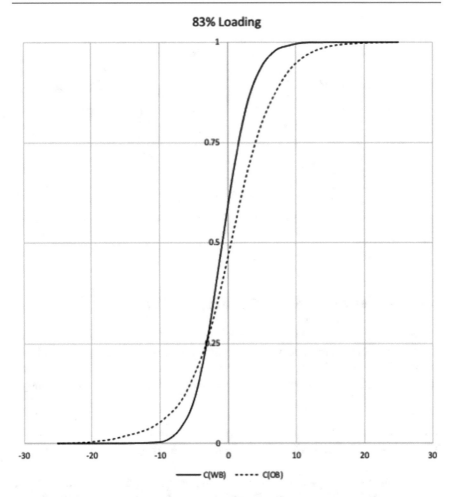

Fig. 9.26 Comparison plot for forecast error distribution at 83% loading

within the production system and would be to use a WB approach (compare the errors plots shown in Figs. 9.18 and 9.21).

The minimum dataset would be the record of job start and finish times at each stage within the production process.

Additionally for either approach there is the need to know the average (expected) processing time at each stage and the distribution of processing times at each stage.

The operations of a larger firm {$C_6[1, 2, 4, 6, 4, 2]$, All Stage Capacities Equal} with loadings of 33%, 67% and 83% were simulated. The plots shown in Figs. 9.26 and 9.27 indicate that as the loading increases a WB system increasingly outperforms the alternative OB system.

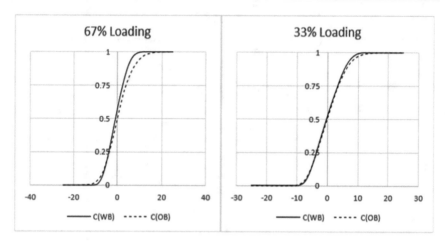

Fig. 9.27   Comparing forecast percentage error distributions

9.6   Data Requirements for Planning and Control

The results from the simulations are summarised in Fig. 9.28 indicating the most appropriate approach to forecasting job completion time at the time of arrival of a new job. The data requirements to enable efficient and effective planning and control, in a Flow Shop, are also dependent on the machine configuration in the workshop.

9.7   Minimal Information in Flow Shops with CONWIP Control

For example considering a constant work in progress (CONWIP), or maximum work in progress (MAXWIP), approach to production control data requirements are centred around the number of jobs present and in a Flow Shop this (initially) implies that a controller will need to know this number before authorising the release of a new job into the system, see Fig. 9.29.

Note: MAXWIP acknowledges the fact that in a workshop 'making to order' on some occasions the number of orders can be less than the level specified for the CONWIP limit. Note: if variability is added at the final stage then work in progress is allowed and a CONWIP production control system is viable but if variability is added at each stage a MAXWIP control system is more appropriate.

Considering first workshops with a single processor at each stage, the configuration within this workshop could cause additional problems, for example if a three-stage workshop had the configuration {NND} queues could form at any stage within this process and much of the work in progress would be caused by jobs waiting for a machine to become free while if the configuration was {NDN} queues could form at the first or second stage with again jobs waiting for a machine to become free.

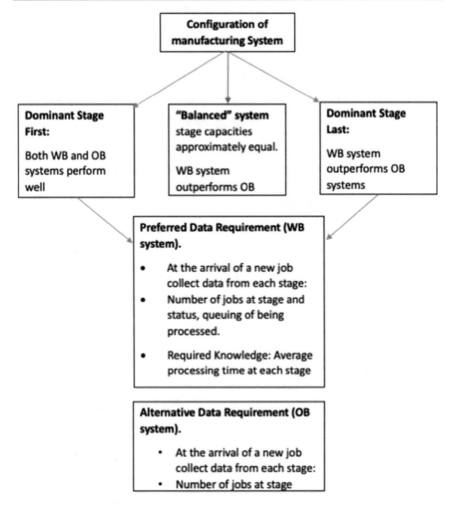

Fig. 9.28 Defining data requirements

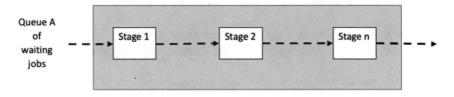

Fig. 9.29 At most, N jobs allowed in the workshop

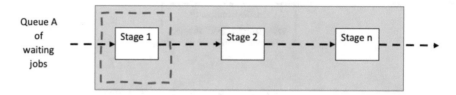

Fig. 9.30   At most, N jobs allowed in the workshop and limit at the first stage

Notice that the configuration {DNN} would minimise the queueing within the workshop, the queue A acting to control the quantity of work in progress and reduce the need for data from within the workshop, the data requirement now being number of jobs entering and number of jobs leaving.

Indicating that the production controller will need to know the configuration of the workshop, in particular the location of the dominant machine, to be able to implement an appropriate, efficient and effective, production planning control system. Thus suggesting that to reduce the data requirement the dominant machine can be used to restrict/allow the entry of a new job into the workshop, when this machine has completed a job a new job may enter the system.

When CONWIP is appropriate information will be required concerning the number of jobs present, number entered less the number leaving the system.

When MAXWIP is employed in addition information will be needed about the number of jobs at each stage, number entering each stage less the number leaving. Also needed would be a MAXWIP1 limiting the number of jobs present at the first stage, required when several jobs arrive simultaneously at the workshop, where the workshop configuration is {NDN} extra monitoring is required at Stage 1, see Fig. 9.30.

More data would be required if the average process times at all stages were the same. Here, information would be needed about the status of jobs at all stages within the production process.

9.7.1   Flow Shops with More Than One Like Processor at Each Stage

If there are n_i processors at stage i and the average processing time is t_i then the values $p_i = \frac{t_i}{n_i}$ can be used to determine the location of the dominant stage in this process.

The minimum data requirement for such a workshop where stage k is the dominant stage would be:

$$N_j = \text{number of jobs up to and including stage } j \text{ for } j \leq k$$

Alternatively in larger workshops the data requirement would be:

$$N_j = \text{number of jobs up to and including stage } j \text{ for all } j$$

9.7.2 Job Shops

These will require more detailed information to enable efficient and effective planning and control to enable an economical use of the production stages.

For example consider a workshop consisting of two stages, stage A and stage B, where a job could follow one of the four paths through the network:

(a) A
(b) B
(c) AB
(d) BA

A possible methodology could be to restrict the number of jobs present to N and those requiring:

- Stage A to n_A
- Stage B to n_B

However, assuming that $n_A = n_B$ then the maximum number of jobs allowed to be present for the following order books:

(e) Jobs of type (a) and (b)
(f) Jobs of type (a) and (c)
(g) Jobs of type (b) and (c)
(h) Jobs of type (c) and (d)

- For case (e) it follows that a total of $n_A + n_B$ jobs could be present
- For case (f) it follows that a total of n_A jobs could be present
- For case (g) it follows that a total of n_B jobs could be present
- For case (h) it follows that a total of n_A jobs could be present

Here the data requirement would, initially, seem to be concerned with the number of jobs at each stage but were the jobs present of types (a) and (d) or of types (b) and (c) then more precise data about the status of each job (proportion of work completed at each stage) will be required to enable control.

A Flow Shop can be controlled through the number of jobs present but as the job routings become more complex addition data concerning the status of each job at each stage will be required to enable efficient and effective planning and control.

9.7.3   Effect of Growth on Planning and Control in a Flow Shop

A firm may grow by:

- increasing the number of stages in the production process;
- increasing the number of processors at each stage in the production process;
- increasing both the number of stages and the number of processors at each stage.

Simulations were constructed to be able to investigate the effect of these possible changes, for each investigated configuration two cases were considered:

- dominant stage at the first stage of the production process;
- dominant stage at the final stage of the production process.

The workshops considered ranged from:

- Small firm—One stage one processor; to
- Large firm—Ten stages and ten processors at each stage.

For each configuration the effectiveness of both linear and piecewise linear models and their associated data requirements were evaluated.

9.7.4   Results from Three-Stage Models

Where the first stage is dominant DNN the most appropriate approach to forecasting is dependent on the firm size and consequently the data requirement will also be dependent upon the size of the firm, see Table 9.7.

Results where the final stage is dominant are given in Table 9.8.

Table 9.7 Results for DNN cases

Forecasting methodology				
	Parameter used		Evaluation	
Number at stage	Number at first stage	Number in system	Best approach	Reliability of approach
1	Linear model	Linear model	Number at first stage	Very high
3	Piecewise linear model	Piecewise linear model	Number at first stage	High
5	Piecewise linear model	Piecewise linear model	Number at first stage	Less reliable
10	Constant time model	Constant time model	Number at first stage	High

Table 9.8 Results for NND cases

Forecasting methodology				
	Parameter used		Evaluation	
Number at stage	Number at first stage	Number in system	Best approach	Reliability of approach
1	Insufficient data	Piecewise linear model	Number in system	High
3	Insufficient data	Piecewise linear model	Number in system	Quite High
5	Piecewise linear model	Piecewise linear model	Number at first stage	Less reliable
10	Constant time model	Constant time model	Number in system	Quite High

9.7.5   Results from 10-Stage Models

Where the first stage is dominant {DN?N} the most appropriate approach to forecasting is dependent on the firm size and consequently the data requirement will also be dependent upon the size of the firm, see Table 9.9.

Results where the final stage is dominant are given in Table 9.10.

These results establish that the location of the dominant stage in a production process is an important factor in the ability of the firm to be able to forecast, plan and control its production processes and that as a firm grows the data requirement to enable efficient and effective control increases.

As the number of processors at each stage increases (describing a large firm) the simulation results show that while forecasting completion time tends to become

Table 9.9 Results for D...N cases

Forecasting methodology				
	Parameter used		Evaluation	
Number at stage	Number at first stage	Number in system	Best approach	Reliability of approach
1	Linear model	Piecewise linear model	Number at first stage	High
3	Piecewise linear model	Piecewise linear model	Number at first stage	Quite High
5	Piecewise linear model	Piecewise linear model	Number at first stage	Less reliable
10	Constant time model	Constant time model	Either	Quite High

Table 9.10 Results for N...D cases

Forecasting methodology				
	Parameter used		Evaluation	
Number at stage	Number at first stage	Number in system	Best approach	Reliability of approach
1	Insufficient data	Piecewise linear model	Number in system	High
3	Insufficient data	Piecewise linear model	Number in system	Quite High
5	Insufficient data	Piecewise linear model	Number at first stage	Not reliable
10	Constant time model	Constant time model	Either	Quite High

simpler/less data requirement effective production control will still need information about the progression of jobs through the system.

A consequence of this result, found in the small manufacturing firm case studies, is that the manager/production planner acts to restrict entry into the workshop thus employing an (informal) CONWIP or MAXWIP system and transforming the more difficult configurations, where the first stage is not dominant, into a production system where the first stage, the manager, is dominant reducing the data requirement for planning and control.

9.8   Conclusion

The data requirement for efficient and effective production planning and control is dependent on the location of the dominant stage in the production process. Without a CONWIP-based control system then:

- When the dominant stage is first (or close to first in a larger process either order-based or work-based) approaches can be appropriate.
- When the dominant stage is towards the end of the process then the most appropriate approach is by way of a work-based system.

Consequently, the minimal information set for a general problem (no conditions regarding the location of the dominant stage) would be the number of jobs at each stage of the production process and the implementation of a work-based system.

When there is a CONWIP-based control system then the best system is dependent upon the size of the firm (number of stages and number of processors at each stage) then recommended data requirements are shown in Tables 9.7 to 9.10.

In particular, for the largest firm the use of a simple fixed time, regardless of work in progress, will lead to the most reliable forecast. Workshop size is an important factor with smaller firms possessing relatively fewer resources experiencing the greatest problem (work needed to estimate delivery times).

9.9 Learning Activities

? Exercise

1. Compare and contrast the information needs of firms of type DN..N, dominant stage first, with those of type NN..ND, dominant stage last.
2. Compare the production planning and control needs of firms of type DN..N with those of type NN..ND.

? Extension exercise

A firm has a manufacturing system consisting of several stages with a single processor at each stage where each stage capacity is greater than that at its previous stage. Within this system queues predominantly occur at the dominant stage.

Explain what the effect on the information requirements will be, to enable planning and control, if there is a need to increase workshop capacity or to reduce production time. Evaluate how this might be affected if the company obtains additional capacity at the dominant stage.

Business Adoption of Analytics   10

10.1 Introduction

Meeting the commercial demands of global manufacturing growth is a ubiquitous challenge for the UK advanced manufacturing sector. Global manufacturing output is projected to grow at a rate of 4.5% by the end of 2018 and remain on an upward trajectory [23]. China is currently leading in terms of global market share, with the UK ranked 6th in the Global Competitive Manufacturing Index.

The UK's position, however, is forecast to drop to 8th by 2020, as Asian countries make progress in manufacturing automation alongside the development of logistically sophisticated, cost competitive, global distribution networks. The UK Government Autumn Budget 2017 Statement [9] states that

> ? to ensure that the next generation can look forward to a better future than the one before it, the Budget: backs innovators who deliver growth; helps businesses create better, higher paying jobs and makes sure everyone has the skills they need to succeed in the new economy.

10.1.1 Intelligent Manufacturing

UK manufacturers are challenged with joining the race to develop 'intelligent manufacturing' facilities. These operations need to be supported by complex and intuitive supply chains that are time responsive and flexible to customer need, while producing precision, high-quality products on a commodity scale, at globally competitive prices.

With the onset of the fourth industrial revolution [20], the UK's potential advantage lies in its lengthy history of precision and high-quality product manufacture. UK manufacturers are aspiring to achieve this growth with 45% indicating they are changing the range of products they will offer, 56% entering new geographical markets over the next 12 months and 49% entering into new sectors [15].

© Springer Nature Switzerland AG 2021
R. Hill and S. Berry, *Guide to Industrial Analytics*, Texts in Computer Science,
https://doi.org/10.1007/978-3-030-79104-9_10

However, the KMPG report also indicated that while a significant proportion of manufacturers recognise the opportunities for digitisation, they 'seem less sure about how it will affect their business and whether they have a coherent strategy and the right talent and skills to capitalise on it'.

10.1.2   Compounded Challenges for SMEs

A large proportion of businesses in the region are SMEs and face challenges emerging from not only the adoption of Industry 4.0 but also the impact of Brexit.

In October 2017, BEIS published the Made Smarter Review Report in readiness for the Industry Strategy White Paper. This report highlights the importance of digital transformation for the future of UK manufacturing and suggests that the UK is 'lagging behind' its competitors, calling for a national drive to raise awareness of digital manufacturing and the potential opportunities for manufacturers in adopting digitally enabling technologies, thus facilitating competitiveness and productivity of supply chains.

Industry 4.0 technologies include automation and data exchange in manufacturing, encompassing cyber-physical systems, the Internet of Things [4], cloud computing, robotics, artificial intelligence (AI) and advanced data analytics and visualisation.

The North of England hosts 28% of UK manufacturing jobs and, with around 12,000 manufacturing companies, the Yorkshire and the Humber region has the second highest number of manufacturing jobs in the UK after the Midlands [18]. At a city-regional level, the Leeds City Region has the largest number of manufacturing jobs in the UK, employing in the region of 140,000 workers [16].

We estimate that there are 6000 manufacturing businesses in the Sheffield and Leeds City Regions that would significantly benefit from digitising their manufacturing capabilities to achieve business growth, increase market share and cumulatively deliver a substantial uplift in regional GVA.

The draft Inclusive Industrial Strategy for the Sheffield City Region states that manufacturing grew by 13% in the period 2010–2015 with the strategic aim to grow more high-quality GVA jobs in this sector [22].

10.1.3   Regional Challenge

There is a predominance of traditional 'craft' manufacturers within the Yorkshire region that will need to respond to these challenges in order to remain competitive. As the momentum of digital manufacturing adoption accelerates, there is a decreased window of opportunity to 'catch-up' by adopting technologies that have existed for some time now.

Such technologies generally relate to the uptake of sales order processing/ Enterprise Resource Planning (ERP) systems, of which many SMEs have so far remained competitive without. Decreasing hardware costs, together with the availability of computing as a utility, means that companies can build systems upon infrastructure that they do not need to own nor manage.

There is an additional challenge for traditional management accounting attitudes, in that such systems move investment from capital expenditure to operational expenditure, which is not always seen as being favourable. We discuss an approach that attempts to overcome some of the challenges of traditionally held beliefs later in the article.

The facilitation of access to external, openly available datasets, does provide massive potential for regional SMEs, and this single example in itself can provide significant impetus for the adoption of digital manufacturing approaches.

Our work has revealed many instances of organisations that are managed by their owner, with combined responsibilities of strategy and operations that need to be balanced according to market pressures. For instance, a business owner may monitor price variability of raw materials and manage their stockholding in response to this.

The availability of cheap, off-premise data storage, together with on-demand high performance computing (HPC) power enables forecasting techniques that can respond in real time to market price changes.

Forecasting and predictive modelling is an example of a relatively mature discipline in the algorithmic sense; the adoption of established techniques that previously required expensive hardware and specific expertise to maintain is now no longer the case, as utility computing together with increased awareness and toolkits that abstract software developers away from the mathematical detail of forecasting become available.

These combined capabilities are now within the reach of SMEs and they can deliver a significant stream of intelligence for a resource constrained enterprise.

10.2 A Model of Engagement

In such a complex landscape, it is challenging to attempt to address the issues faced by both industry and academic institutions. The university was commissioned by the Digital Catapult to undertake an investigation to understand the appetite and readiness for adoption of Industry 4.0 amongst manufacturers in the Yorkshire Region. A total of 51 organisations (26 manufacturers, 6 digital services to manufacturing and 19 regional/national trade associations and membership representative bodies) were interviewed.

A report on the findings of the study was subsequently published by the Digital Catapult 'The Future of Manufacturing in the Digital Age' [5]. The report outlined a mixed picture with a larger than expected volume of companies who could best be described as digital laggards or cynics. There was, however, a cohort of companies who could be described as being cautiously positive about Industry 4.0 (a group which became known as *cautious innovators*).

10.2.1 Proving the Return on Investment

A number of the cautious innovator stakeholders were keen to explore the adoption of industrial digital technologies (IDT) in more detail. These firms had in common a desire to explore how IDT could add value to their business; however, they were unclear regarding: potential Return of Investment (ROI); which technologies they should adopt first; or how to fully implement these changes in their companies.

These manufacturers saw the benefit of becoming more digitally enabled or 'smarter' and recognised that there was plenty of scope to do this by making more value from the data they had already collected. These firms generally wanted to find ways to undertake Proof of Concept (PoC) activities to enable them to gauge the possible Return On Investment (ROI) relating to better data analytics, and move from using data to monitor processes, to harnessing the predictive capabilities of data, could have a material impact on their bottom line.

The University of Huddersfield is an active academic partner of the Digital Cata-pult and became actively involved with events and initiatives run by the Catapult at a regional and national level. As part of this, the university was asked to be a judge on a number of industrial competition events or 'Hack and Pitches'. This brought the university into contact with a wide range of specialist IDT SMEs mainly in the field of data analytics and artificial intelligence (AI). Many of these firms were seeking to gain market exposure in manufacturing.

In discussion, it was considered by many of these, mainly start-up companies that they found it hard to access manufacturers and in particular SMEs who, it was felt, may be more receptive to try novel and small scale and industrial digital technology (IDT) innovations. These digital start-ups were often located in major cities, in particular London (improved access to venture capital and investment as well as analytical skill sets), but were detached from the major centres of manufacturing such as West Yorkshire.

A plan was developed whereby all the stakeholders involved were able to extract value (the manufacturer, the digital technology company and the university). At a practical level, the university acts as the facilitator and honest broker to introduce specialist digital technology companies to the manufacturers.

The start-up digital technology companies are given the opportunity to prove their services to an industrial partner and gain a market presence, and the manufacturer is given the much needed PoC and ROI they are looking for. At a strategic level, the university is able to forge an underpinning and trusted platform for a more comprehensive strategic partnership with the companies moving forward.

10.2.2 Digital Enablers Network (DEN)

The university initiated a series of practical introductions and ongoing dialogues between manufacturers, the university and innovative digital technology providers (or 'digital enablers' as they are referred to). These dialogues with the digital enablers have resulted, in a number of cases, to a commitment to develop further PoC activities.

The university subsequently decided to formalise this approach by establishing the Digital Enablers Network (DEN) model, which offers digital technology SMEs and start-ups access to the manufacturer, with the university acting as trusted expert advisor.

In turn, the manufacturers are able to access to specialist small scale Proof of Concept type projects, often on a low cost or pro-bono basis, as a pre-cursor for more substantial investments over time. This three-way partnership between the manufacturer, the digital enabler and academia has flourished and has led to demonstrable impact.

In parallel with this activity, academics were also conducting research into myriad technologies that facilitate the desire to link cyber-physical systems. This work was evaluated in terms of its practical applicability for the cautious innovator group, and research that directly supported the likely issues that a first-time adopter of digital manufacturing technologies might face, such as network bandwidth saturation, whether to process sensor data at the edge of the network, scalable architectures, etc. [7,10,17,21].

From the intelligence that was gathered about the region, set against a backdrop of the UK industrial strategy, we summarise the following characteristics of the approach:

- Response. SMEs need a responsive service. The agility required is not always serviceable from a university, where there is typically the expertise but not the necessary capacity;
- Research-based. The approach needs to be evidence-based and itself responsive to the varying needs and domains within the region;
- Multi-disciplinary. Key stakeholders—cautious innovators—are likely to be the most receptive audience. Nonetheless, they will require more than just technical assistance, and help with the adoption and management of change will be welcome. Hence, the approach needs to have multi-disciplinary skills at its disposal;
- Relevant. Similarly, if the changes are to be adopted and embedded into organisations that are focused on ROI, the case for adoption must be couched/presented in the same currency accordingly;
- Trustworthy. Again, to reflect the traits of the target audience, the approach must actively promote and build trust between the cautious innovators, the niche supplier and the university. This would be achieved through Proof of Concept work that demonstrates/visualises value to the business in situ with their data.

We have chosen to develop a model of engagement that incorporates these characteristics, illustrated in Fig. 10.1. Our approach is to put all of the eggs in a basket, and then 'watch the basket'. In this case, the University, via CIndA, provides an environment that supports the incubation of a multi-faceted set of relationships.

This invariably consists of the following services: client/DEN member matching (as well as associated selection and filtering); creating business cases; market penetration; research expertise; skills development; ROI evaluation and impact assessment; and funding advice and partnership.

Fig. 10.1 DEN model and stakeholders

10.3   University Capability

In January 2017, the University of Huddersfield secured funds to establish a Centre for Industrial Analytics (CIndA). One of the main challenges faced was that although the Centre for Industrial Analytics was created with domain expertise and capability, it was early in its development and lacked capacity to respond to the high level of demand from industry.

It also lacked a track record of delivery with industry, and there was a pressing need to provide practical and immediate benefits for businesses, in particular busy manufacturing SMEs. As described in earlier, the Digital Enablers Network was conceived as part of a structured model to address the challenge.

10.3.1   Case Study: DEN in Action

Flexciton (FL) is an agile data analytics/artificial intelligence start-up company, mentored and supported by the University of Huddersfield, based in London. They have developed bespoke AI software capabilities designed to create productivity and efficiency gains in industry, through optimised scheduling and work planning processes.

The CEO of FL presented at a Digital Catapult Hack and Pitch event, which was attended by the university. Flexciton demonstrated capability, ethos and a willingness to work with industry and academia and felt it necessary to be based in London in order to access venture capital investment and specialist skills. Conversely, London is some distance from the manufacturing marketplace they wanted to operate in, and they had little or no connections into the industry. CF is an innovative textile company based in West Yorkshire. They are an important part of the UK textile supply chain, producing specialist cloth, mainly for the furnishing in the transportation sector, and are at the centre of a large and complex supply chain. CF produces 150 km of cloth per week, from a wide catalogue of its own products and designs.

Representatives from CIndA met with senior managers in the plant and identified refinements to their capabilities which might have a positive impact on their production, in particular, workload planning and scheduling. The university facilitated and mentored both parties through every step of the process, acting as a trusted advisor and providing objective domain expertise. An agreement was brokered by all parties for FL to undertake a PoC and CF promptly released samples of production related data for the FL to analyse.

FL were able to concentrate time and expertise to the challenge and within a couple of weeks presented their findings and demonstrated the use of their core software capability. FL fed back to CF and through the use of persuasive data visualisation techniques were able to articulate a clear business case for action, demonstrating with a high degree of accuracy precisely how a plethora of minute adjustments to work planning and scheduling could make substantial production gains for the company.

These insights had a tangible and immediate impact upon CF senior managers, who up to that point would have been fully justified in being guarded and cautious about the merits of working with a data analytics company such as FL. A clear level of trust and mutual belief had been engendered by all parties. CF and FL then entered a period of negotiation looking for an agreement for CF to fund the next phase of implementation.

The university again acted as the facilitator of these negotiations in the process, offering assurance and expert opinion as required. CF chose to progress and invested in the next stage of the project, enabling FL to move to implementation.

The university continued to offer overarching support coaching and mentoring across the partnership. Working with all parties, the university enabled a wider discussion around an ambitious strategy for a full scale, enterprise-wide program of digital transformation for the company.

Working closely with the CF and FL, the CIndA was able to develop a roadmap for the adoption of Industrial Digital Technologies (IDT), including the introduction of novel Industrial Internet of Things applications (IIoT) and other emerging technology solutions such as Microservices Architectures [7,21], with inspiration from prior work in distributed community healthcare provision [8] across their whole connected supply chain, further utilising the capabilities of the university and the wider DEN network.

10.4 Discussion

When recruiting a potential DEN member, it is sometimes necessary to explain what we see as a need for pro-bono work upfront before a 'sale' is made. While analytics consulting can be a high value/return activity, trust needs to be engendered with a client to ensure that the work will deliver value. The success is not all about analytics; that alone can at best only answer one question at a time.

It needs to be combined with additional sensing, automation, appropriate visualisation (for comprehension) and sensitive change management/technology adoption. An important part of this is capability building within the client organisation, at all levels. Data literacy and analytics skills have already been highlighted by many as a key strategic response to the growth of digital manufacturing.

10.4.1 Benefits to SMEs

For an SME, any future vision of digital manufacturing needs in part to be made concrete. This is best done by proving the value of the capability with the organisation's data. Illustrating the power and speed of analysis leads to shared enlightenment, followed by new, more relevant, better-formed questions being posed based on evidence.

The value to an SME is best demonstrated by (a) confirming something the business already knows, (b) identifying something the business did not know and (c) providing the evidence for something the business always suspected but could not prove.

10.4.2 Benefits to den Members

DEN members are mostly start-ups seeking case studies to demonstrate their capabilities or established companies seeking to penetrate new markets, who therefore want case studies. The more cases they have, the greater the experience they have to draw upon when selling products or services to potential customers.

10.4.3 Benefits to Academia

This activity directly supports the impact agenda, including the fostering of closer ties between academia and industry. Real-world evaluation of technologies and approaches to adoption creates robust datasets for subsequent analysis. Such work is inherently multi-disciplinary: business, organisational, operations, psychology; this enables the university to interface more frequently with industry, while also broadening the research experience of researchers in traditionally separate domains.

In addition, CIndA is also a vehicle through which existing research can be disseminated more directly into industry for greater impact. For instance, the Department of Computer Science, working with CERN, is investigating solutions to high velocity edge analytics, which are not dissimilar to emerging challenges in industry.

10.4.4   Human Factors

Human factors are critical to the successful adoption of digital manufacturing approaches for a number of reasons. First, the analysis of data is relatively straight-forward and formulaic, but it is the comprehension of data that is predominantly a challenge. Visualisation is the key here, but as data streams become more complex, the demands for more sophisticated visualisation approaches (e.g. augmented reality/immersion/digital twinning) become more apparent.

Second, the underlying technologies of digital manufacturing are not always perceived as being directly relevant to the problem to be solved. For example, the introduction of additional sensing is a tangible way of providing more 'visibility' of a process. A Hadoop cluster, or an Amazon Web Service instance on which forecasting models are to be executed, is seen as more abstract and therefore more difficult to visualise.

Third, and potentially the most damaging, is the effects upon existing staff feeling threatened by the introduction of technologies which might result in them being no longer required, or at least their positions of power being exposed. A perception that this is a potential issue, together with a lack of confidence from managers as to how the change can be managed, can lead to reticent behaviour even in the light of a strong financial justification to adopt the change. Any adoption project needs to consider the human factors and incorporate change coaching as an essential and intrinsic part of the process.

The university's ultimate role is to engage in knowledge transfer. Our model makes progress towards this and exploits the multi-disciplinary capability that is unique to higher education institutions. However, this model is also demonstrating how a university can move beyond the transactional behaviours of knowledge transfer to being the provider of a knowledge platform, where multiple parties can come together to realise shared value.

10.5   Conclusions

We attribute the success of the engagement to the following four factors:

1. Direct engagement of university head of business engagement with academic head of department. Many higher education institutions struggle to manage the interface between academia and enterprise.
2. Direct recruitment of private companies to deliver technological innovation that already exists within the university. Universities are resourced and planned around teaching requirements, with a cycle of workload that is understood throughout the year. The on-demand capacity requirements of ad hoc projects cannot always be catered for by an academic department, whereas a small business start-up has the flexibility and agility to deliver response to a business. Consequently, the university can focus on match-making, governance, Quality Assurance, impact

assessment and evaluation and mentoring of both industrial partners, which is inherently more development focused.

3. The creation of a multi-disciplinary capability is essential. The successful adoption of digital manufacturing is not solely concerned with cloud computing, predictive analytics or High Performance Computing; to a greater or lesser extent these technologies are essential components of a much broader adoption approach that challenges the ways in which organisations work and collaborate. This requires expertise across many of the disciplines that traditionally have been separated within academia, thus requiring a multi-disciplinary attitude to make the most of the opportunities.

4. The university's role as the architect and governor of the value exchange platform. ROI is the primary assessment mechanism in use by SMEs, and it is an imperative that the university understands this, while interfacing with external funding schemes that may utilise other measures.

10.6 Future Work

We feel that there are a number of pertinent issues that CIndA must explore to further develop our successful model of engagement and are therefore open issues for the research community.

First, there is a need for more practical demonstrations of the retrofitting of technology to existing plant, e.g. LoraWAN, edge analytics [6,7], wireless sensor networks, secure use of hybrid and public clouds [2], software-defined networking/network function virtualisation, etc. It is the development of robust methods for deploying such technologies within real-world settings that will illustrate the potential of advanced data analytics and visualisation for data comprehension.

Second, more cases of action research-based work around the coaching of attitudes for technology acceptance/adoption. Digital manufacturing technologies make greater demands on the people skills (emotional intelligence) of staff who have not been required to do this in their pre-existing organisational 'silos'.

Finally, cases of companies that use data and networking to break traditional physical boundaries and share knowledge for the common good will help develop intelligence around the shared use of data amongst manufacturing companies. For instance, adopting peer-to-peer (P2P) social networking approaches for the sharing of process data for improved coordination [3,8,19] and enhanced efficiencies/profitability [1].

Many smaller organisations already work in regional communities, especially in the B2B environment. Indeed, more work to explore the benefits of closer integration of systems, while also implementing looser coupling (enabling businesses to choose from a greater selection of partners, perhaps working with specialists), will ultimately lead towards models of distributed manufacturing.

We see that the common thread in the challenges faced by regional manufacturers are firmly rooted in the human factors of adoption, whether it be data comprehension, technology adoption or a mindset that embraces business models that are enabled by the innovative collaboration possibilities of shared data.

References

1. Al-Aqrabi H, Liu L, Hill R, Antonopoulos N (2012) Taking the business intelligence to the clouds. In: 9th international conference on embedded software and systems (HPCC-ICESS). Liverpool, IEEE Computer Society, pp 953–958
2. Al-Aqrabi H, Hill R (2018) Dynamic multiparty authentication of data analytics services within cloud environments. In: 2018 IEEE 20th International Conference on High Performance Computing and Communications. In: IEEE 16th international conference on smart city; IEEE 4th international conference on data science and systems (HPCC/SmartCity/DSS), pp 742–749
3. Beer M, Huang W, Hill R (2003) Designing community care systems with AUML. IEEE Computer Society
4. Bessis N, Xhafa F, Varvarigou D, Hill R, Li M (2013) Internet of things and inter-cooperative computational technologies for collective intelligence. In: Studies in computational intelligence, vol 460, Springer, Berlin. ISBN 978-3-642-34951-5
5. Digital Catapult (2017) The Future of Manufacturing in the Digital Age, August 2017, https://www.digitalcatapultcentre.org.uk. Last accessed 30 Jan 2017
6. Garcia-Campos J, Reina D, Toral S, Bessis N, Barrero F, Asimakopoulou E, Hill R (2015) Performance evaluation of reactive routing protocols for VANETs in urban scenarios following good simulation practices. In: 9th international conference on innovative mobile and internet services in ubiquitous computing, pp 1–8
7. Hill R, Devitt J, Anjum A, Ali M (2017) Towards in-transit analytics for Industry 4.0. FCST2017. IEEE Computer Society, Exeter
8. Hill R, Polovina S, Beer M (2005) From concepts to agents: towards a framework for multi-agent system modelling. In: Proceedings of the fourth international joint conference on Autonomous agents and multiagent systems (AAMAS '05). ACM, New York, NY, USA, pp 1155–1156. https://doi.org/10.1145/1082473.1082670
9. HM Treasury (2017) UK Government Autumn Budget 2017 Statement, HM Treasury. https://www.gov.uk/government/publications/autumn-budget-2017-documents/autumn-budget-2017. Accessed 30 Jan 2017
10. Ikram A, Anjum A, Hill R, Antonopoulos N, Liu L, Sotiriadis S (2015) Approaching the Internet of things (IoT): a modelling, analysis and abstraction framework. Concurrency Comput Practice Exp 27(8):1966–1984
11. ITA (2020) The Institut fur Textiltechnik (ITA) of RWTH Aachen University
12. McKinsey and Company, Inc (2016) McKinsey Industry 4.0 Global Expert Survey, New York
13. Mistry P, Lane P, Allen P, Al-Aqrabi H, Hill R (2019) Condition monitoring of motorised devices for smart infrastructure capabilities. In: International conference on smart city and informatization, Springer, Berlin, pp 392–403
14. Kusters D, Pra N, Gloya YS (2017) Textile learning Factory 4.0 preparing Germany's textile industry for the digital future. In: 7th conference on learning factories, CLF 2017—procedia manufacturing, vol 9, pp 214–221
15. KPMG (2017) Global Manufacturing Outlook Survey 2016
16. Leeds City Region (2016) Strategic Economic Plan 2016–2036 http://investleedscityregion.com/system/files/uploaded_files/SEP-2016-FINAL.pdf. Accessed 30 Jan 2017

17. Newman S (2017) Building microservices—designing fine-grained systems. O'Reilly
18. ONS (2016) UK Business Register and Employment Survey (BRES)
19. Polovina S, Hill R (2005) Enhancing the initial requirements capture of multi-agent systems through conceptual graphs. In: International conference on conceptual structures. Springer, Berlin, pp 439–452
20. PWC (2016) Industry 4.0: building the digital enterprise. http://www.pwc.com/industry40. Accessed 30 Jan 2017
21. Shadija D, Rezai M, Hill R (2017) Towards an understanding of microservices. In: Proceedings of the 23rd international conference of automation and computing (ICAC). University of Huddersfield, IEEE Computer Society
22. Sheffield City Region (2015) Strategic Economic Plan 2015–2025. https://sheffieldcityregion. org.uk/wp-content/uploads/2018/01/SCR-Growth-Plan-March-2014-1.pdf. Accessed 30 Jan 2017
23. World Bank (2017) Global Outlook, Global Economic Prospects. http://pubdocs.worldbank. org/en/216941493655495719/Global-Economic-Prospects-June-2017-Global-Outlook.pdf. Accessed 30 Jan 2017

Statistics

This appendix contains reference materials for use in conjunction with the learning exercises within the main text.

A.1 Basic Descriptive Statistics

A.1.1 Measures of Location

- *Mean*—used to represent the typical job times. the mean of n items x_i is given by:

$$\mu = \frac{\sum x_i}{n} \tag{A.1}$$

- *Modal value*—the most commonly occurring value. Can be used in place of the mean.
- *Median*—the central value in a set of values. When all of the job times are placed in ascending order then the median value is the *value in the centre*.

A.1.2 Measures of Spread

- standard deviation (SD);

$$\sigma = \sqrt{\frac{\sum (x_i - \mu)^2}{n}} \tag{A.2}$$

- mean absolute deviation (MAD);

$$MAD = \frac{\sum |(x_i - \mu)|}{n} \tag{A.3}$$

- coefficient of variation: this expresses the variability in the data.

$$CV = \frac{\sigma}{\mu} \tag{A.4}$$

© Springer Nature Switzerland AG 2021
R. Hill and S. Berry, *Guide to Industrial Analytics*, Texts in Computer Science,
https://doi.org/10.1007/978-3-030-79104-9

A.2   Statistical Distributions

A.2.1   Describing Job Arrivals

The exponential distribution describes those applications where the intervals between the arrival times of jobs is random, with a known average time between successive arrivals.

Often used to describe the input into a workshop where orders arrive as single jobs, for example a small production facility.

Use in Simulations: If k is the average time between successive arrivals then the probability of an arrival by time T is given by:

$$F(T) = 1 - e^{-LT}, L = \frac{1}{k} \tag{A.5}$$

$$rand = 1 - e^{-LT}, 0 \le rand \le 1 \tag{A.6}$$

$$T = -\frac{ln(1 - rand)}{L} = -kln(1 - rand) \tag{A.7}$$

The Poisson distribution describes those applications where a random number of jobs arriving a day is random with a known average number of arrivals per day and often used to describe the number of jobs per day at a large furniture manufacturing firm where customers will order multiple items.

The probability of x arrivals, where the mean number of arrivals per day is m, is given by:

$$p(x) = \frac{e^{-m}m^x}{x!} \tag{A.8}$$

Use in simulations: The cumulative probability (0 to n arrivals) is given by:

$$P(n) = \sum \frac{e^{-m}m^x}{x!}, x = 0, 1, \ldots, n \tag{A.9}$$

To generate the number of arrivals find n so that for a random number R:

$$P(n) \le R \text{ and } P(n - 1) \ge R \tag{A.10}$$

where R is a random number such that $0 \le R \le 1$.

A.2.2   Describing Job Times

When a task is repeated (many times) it is normally assumed that the job times are described by a normal distribution.

However, when the parameters for this distribution are not known (as in a workshop) approximate distributions can be used to describe the data.

Deterministic: when there is no real data it can be assumed that the mean (or modal or median value) describes the processing times, all times equal thus $\sigma = 0$.

Table A.1 Sample data

Sample size	Values	Max	Min	Sample mean	Mean estimate	sd estimate
10	101.3335	105.0271	82.24054	95.34322	93.63383	3.797764
25	85.98078	124.8206	80.6258	100.7099	102.7232	7.365798
50	85.47104	123.8174	77.22188	97.22719	100.5196	7.765924

Uniform (or rectangular) distribution: when more information is available job times can be assumed to be within a given range, with all times equally likely, this distribution is suitable when little information exists with respect to process times.

The only information needed is 'estimates' for the most optimistic and most pessimistic durations.

$$\{\text{Min}, \text{Max}\} = \{\mu - a, \mu + a\}, \text{where Mean} = \mu, sd = \frac{a}{\sqrt{3}} \tag{A.11}$$

Triangular distribution: when much more data is available, typically the most likely job duration in addition to the minimum and maximum times, giving the distribution:

$$\{\text{Min}, \text{Mode}, \text{Max}\} = \{\mu - a, \mu, \mu + a\}, \text{where Mean} = \mu, sd = \frac{a}{\sqrt{6}} \tag{A.12}$$

If the actual mean process time and standard deviation of the process times are known, much data available, μ and σ then job times can be modelled by a normal distribution.

Note: For comparison if the maximum and minimum observed values are assumed to be 6σ limits for the distribution then the parameters for a normal distribution could be estimated. Giving the estimates:

$$\text{Mean estimate} = \frac{(\text{Maximum} + \text{Minimum})}{2} \tag{A.13}$$

$$sd \text{ estimate} = \frac{(\text{Maximum} - \text{Minimum})}{6} \tag{A.14}$$

Example: Values were sampled from a normal distribution with mean (μ) = 100 and sd (σ) = 10 giving the results: These approximations for the distribution parameters can be compared through the plots of the implied cumulative probabilities (Fig. A.1 and Table A.1).

Likewise comparing the cumulative probability plots for:

The underlying normal distribution and the simpler distributions derived using the maximum and minimum observed values:

- the uniform distribution;
- the triangular distribution.

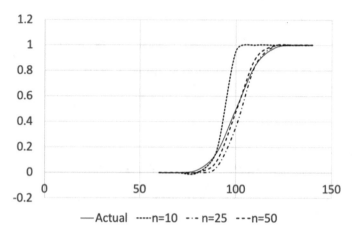

Fig. A.1 Plot comparing estimated values with actual

Where the maximum and minimum values were obtained using 75 (Fig. A.2) and 10 (Fig. A.3) data points. The plots suggesting that a triangular distribution provides a good approximation to the (actual) normal distribution when there is sufficient data to provide reliable estimates for the maximum and minimum values.

A.2.3   Adding Distributions

If in production line all jobs move simultaneously the rate of the 'line' will be determined by the *slowest* (dominant) stage.

Simulating the operation of a production line with five stages where the process times at the dominant stage are described by a uniform distribution (duration 10 to 15) for 100 jobs gave a mean time in the system of 62.1 with a standard deviation of 3.36.

Comparing these results with those generated from a normal distribution with the same mean and standard deviation (Fig. A.4) showed that these times are approximately normally distributed and planners could use this distribution to estimate completion times.

A.2.4   Regression Methods

Used to investigate relationships between two variables, for example number of jobs in the system at the arrival of a new job and time to finish the new job.

Here investigating whether time can be forecast knowing the distance travelled First plot a scatter diagram.

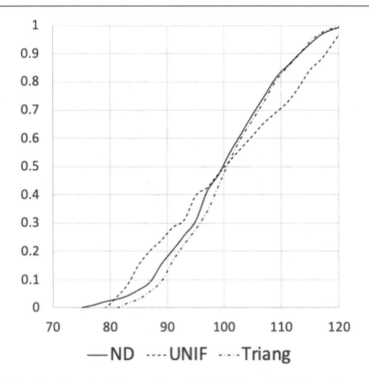

Fig. A.2 Samples of normal, uniform and triangular distributions with 75 data points

Test 1: Does the plot look (reasonably) linear, here YES so continue to determine the equation describing this relationship and associated correlation coefficient (Fig. A.5).

Test 2: Correlation coefficient this can indicate the appropriateness of a linear model (given that Test 1 has been passed):

- Correlation = +1, suggests perfect Linear Trend positive slope;
- Correlation = 0, no evidence of a Linear Trend;
- Correlation = −1, suggests perfect Linear Trend positive slope.

Here:

- Test 1, PASSED
- Test 2, $R = 0.993781$, PASSED close to +1
- Relationship, Time = $8.2037 + 4.9339 \times$ Distance

The model is such that the error:

$$E = \sum (\text{forecast y} - \text{actual y})^2, \text{ is minimised.} \tag{A.15}$$

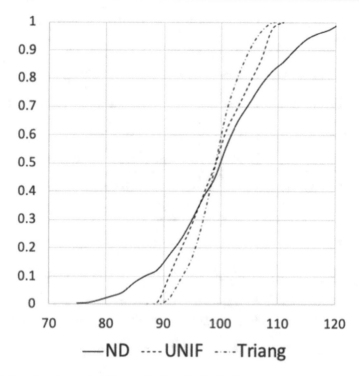

Fig. A.3 Samples of normal, uniform and triangular distributions with 10 data points

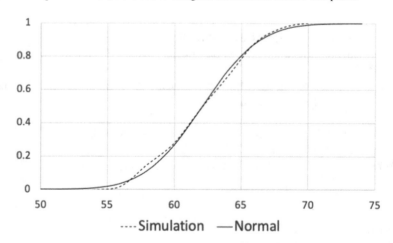

Fig. A.4 Probability plot of times from a five-stage production line

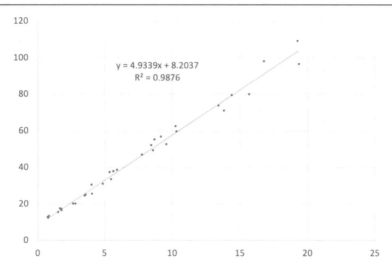

Fig. A.5 Distance against time as a scatter plot

Notice that if a second (alternative) model is required from this data, to forecast distance (y) given travel time (x) start with a new plot with reversed axes: Here:

- Test 1, PASSED
- Test 2, $R = 0.993781$, PASSED close to +1
- Relationship, Distance $= -1.5508 + 0.2002 \times$ Time

Notice that this is not a simple rearrangement of the previously determined relationship Here take the first relationship and dividing through the equation (Fig. A.6).

$$\text{Time} = 8.2037 + 4.9339 \times \text{ Distance or Distance} = -1.662721 \times \text{ Time}$$

A.2.5 Common Failure Cases in Regression

A.2.5.1 Mixed Data

The plot shows that there are two grouping, these should have been investigated independently:

- TEST 1, FAILED

However had the analysis continued, it would have produced:

- TEST 2, Correlation, 0.968917, Close to 1
- Model, $y = 1.0689x + 4.918$

But in effect it is simply joining TWO points together (Fig. A.7).

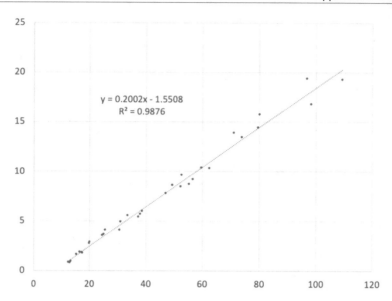

Fig. A.6 Distance against time as a scatter plot

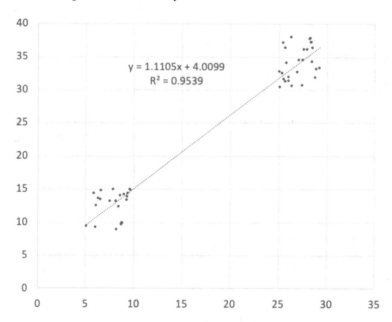

Fig. A.7 Scatter plot showing regression line joining two clusters of data points

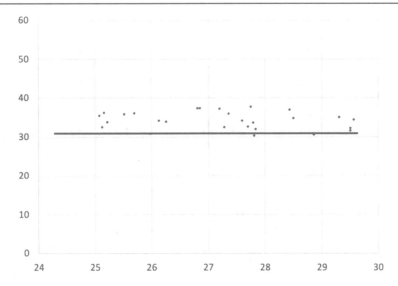

Fig. A.8 Regression line showing zero gradient

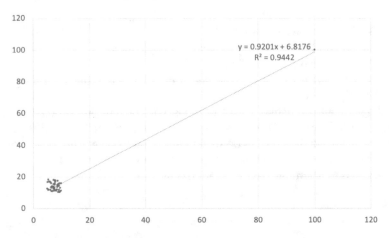

Fig. A.9 Effect of an extreme value (outlier)

Plot the two groups independently. For example plotting the second of these grouping gave no evidence of a Linear Trend (slope ZERO, Fig. A.8).

A.2.5.2   A Point Different From the Rest

For example an organisation with many similar outlets and one large outlet Correlation = 0.9717 (Fig. A.9).

However plotting the common values (excluding the extreme value) gives no evidence of a Linear Trend, as in Fig. A.10.

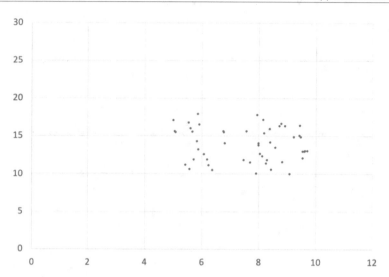

Fig. A.10 Outlier removed from the plot

A.2.5.3 TREND Exists But It Is Not Linear

Using data generated from a workshop with configuration

$\{C3[1, 1, 1], DNN\}$ TEST 1 Failure

But the correlation and regression formulae give an incorrect model (it is a quadratic model) (Fig. A.11).

Fitting non-differentiable (linear) functions. For example, production data shown in Fig. A.12. Where point P is known and point R the location of the non-differentiable point is known and the gradient of the line S is known (knowledge occurring with production data here there are three stages in the production process)

The aim is to determine the two relationships:

$$y = a + bx, x \leq 3$$
$$y = A + Bx, x \geq 3$$

Using the known points and process data (total average production time = 21) and the average time for the dominant stage is 8 gives:

$$\text{At point P } x = 0, y = 21$$
$$\text{At point R } a + 3b = A + 3B$$

For line S, $B = 8$ (average dominant time/number of processors at stage = 8/1 = 8)
Combining gives this data and information leads to a relationship between A and b

$$21 + 3b = A + 3 \times 8 \rightarrow A = 3b - 3$$

The regression lines can now be determined by minimising the error expression (only one variable b)

$$E = \Sigma_{n_1}(y - 21 - bx)^2 + \Sigma_{n_2}(y - 3b + 3 - 8x)^2 \tag{A.16}$$

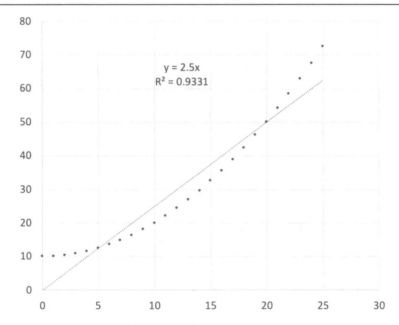

Fig. A.11 Linear model applied to nonlinear data

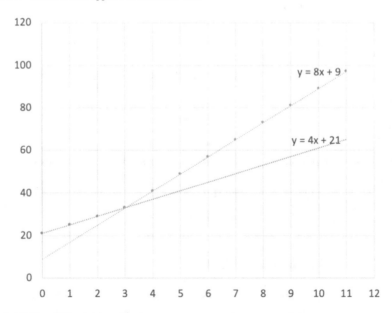

Fig. A.12 Non-differential perfect data

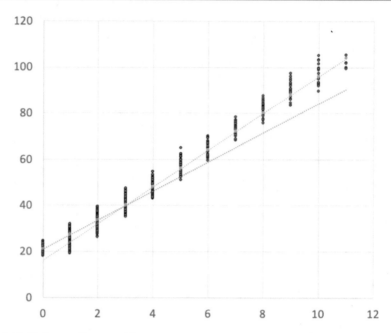

Fig. A.13 Fitting two linear models

Differentiating leads to:

$$\Sigma(y - 21 - bx)x + 3\Sigma(y - 3b + 3 - 8x) = 0 \qquad (A.17)$$

Giving the expression for b:

$$b = \frac{\Sigma_{n_1}(xy - 21x) + \Sigma_{n_2}(3y + 9 - 24x)}{\Sigma_{n_1}x^2 + \Sigma_{n_2}9} \qquad (A.18)$$

Using the data implied in Fig. A.12 leads to the result that $b = 4$, thus giving the model:

$$y = 21 + 4x, x \leq 3$$
$$y = 9 + 8x, x \geq 3$$

A.2.6 Modelling Simulation Data

As a second example data generated (DNN 66% loading) in section XX was investigated using this non-differential function model the results are shown in Fig. A.13 the linear models are:

$$y = 21 + 6.33x, x \leq 3$$
$$y = 16 + 8x, x \geq 3$$

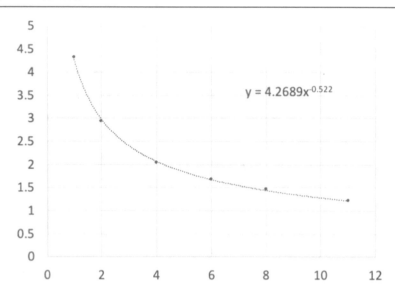

Fig. A.14 Modelling the gradients

In general where there are n stages, let:

$$K = \Sigma \overline{T}_i \text{ and } M = \max(\overline{T}_i) \qquad (A.19)$$

then:

$$b = \frac{\Sigma_{n_1}(xy - Kx) + \Sigma_{n_2}(ny + Mn^2 - nK - Mnx)}{\Sigma_{n_2}x^2 + \Sigma_{n_2}n^2} \qquad (A.20)$$

$$A = nb + K - Mn \qquad (A.21)$$

NOTE (to be checked) in a larger system, many processors at each stage, n = total number of processors.

Two-stage models for an n stage system with a single processor at each stage are shown in table B2 <<INSERT TABLE HERE>> The results for Model 1 are presented in graph B3.10 and this indicates a relationship between the number of stages and the derived gradients. These results indicate that for Model 1 (Fig. A.14)

$$T = A + Bx$$

The values for A can be modelled by:

$$A = 1.5 + 6.5N_{sys}, \text{ and}$$

the values for B can be modelled by:

$$B = 4.27N_{sys}^{-0.52}$$

Simulation Library—Ciw

B

B.1 About Ciw

Ciw (Welsh for 'queue') is an open source queueing simulation library for the Python programming language.

It simplifies the construction of systems that can be modelled using queues. All process-based systems can be represented using queues.

Some of the examples in this book use Ciw to help understand business systems, and the focus is to get to analysing the problem with the minimum amount of programming knowledge.

Details of the library can be found here:

https://github.com/CiwPython/Ciw.

Online documentation (Fig. B.1) can be found at:

https://ciw.readthedocs.io/en/latest/.

B.2 Installing Ciw

You need to have a working installation of Python before Ciw can be installed.

The easiest way to install Ciw is using 'pip' as per the instructions here:

https://ciw.readthedocs.io/en/latest/installation.html.

© Springer Nature Switzerland AG 2021 243
R. Hill and S. Berry, *Guide to Industrial Analytics*, Texts in Computer Science,
https://doi.org/10.1007/978-3-030-79104-9

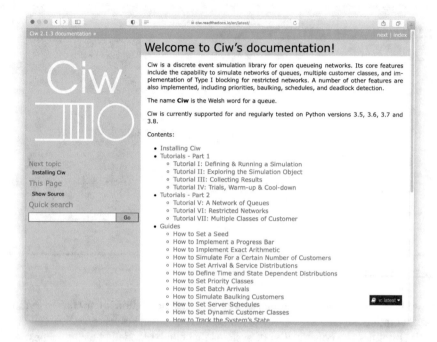

Fig. B.1 Online documentation for Ciw

Production Planning Programs in MS Excel

Two programs in MS Excel can be used to allocate production resources to satisfy, or set, delivery dates for a new order.

- Program 1: This checks whether or not there is sufficient capacity within the system to be able to satisfy a customers desired delivery date, or to establish a possible delivery date.
- Program 2: This enables time to be allocated at each production stage to deliver goods on time to the customer and validating the chosen delivery dates.

C.1 Program 1: Checking Order Delivery Date

C.1.1 Set Up Production Stage Data

For example, given the data for Product A in Fig. C.1: The process times for each stage are as shown in Table C.1.

C.1.2 Set Up Stage Characteristics/Capacities

Set the stage capacities, both normal working and overtime working, For example:

Fig. C.1 Three workstation production line

© Springer Nature Switzerland AG 2021
R. Hill and S. Berry, *Guide to Industrial Analytics*, Texts in Computer Science,
https://doi.org/10.1007/978-3-030-79104-9

Table C.1 Times at each stage for each unit of Product A

	Product A		
	Stage		
	1	2	3
Time	5	2	3

	TIME/ Time Units																		
	1	2	3	4	5	6	7	8	9	10	11	12	13	14	15	16	17	18	19
CAPACITIES NORMAL TIME																			
LINE 1	5	5	5	5	5	5	5	5	5	5	5	5	5	5	5	5	5	5	5
LINE 2	4	4	4	4	4	4	4	4	4	4	4	4	4	4	4	4	4	4	4
LINE 3	2	2	2	2	2	2	2	2	2	2	2	2	2	2	2	2	2	2	2
CAPACITIES OVERTIME																			
LINE 1	1	1	1	1	1	1	1	1	1	1	1	1	1	1	1	1	1	1	1
LINE 2	1	1	1	1	1	1	1	1	1	1	1	1	1	1	1	1	1	1	1
LINE 3	1	1	1	1	1	1	1	1	1	1	1	1	1	1	1	1	1	1	1

Fig. C.2 Loading capacity data for workstations

																NOW	
DATES	1	2	3	4	5	6	7	8	9	10	11	12	13	14	15	16	17	18
																	1	

Fig. C.3 Loading dates into worksheet

LINE 1 (Product A) has a capacity of 5 during normal working time and an additional 1 during overtime working (note these are not necessarily all equal), as per Fig. C.2.

C.1.3   Updated Working Sheets

Sheets TEMPLATEFULL and DandL will have now been updated using this data.

C.1.4   Input Date and Know (Feasible Order)

On worksheet COMPLOTS Set the current TIME/DATE. If the date is ?time=17? enter a 1 into date cell 17 (Fig. C.3). Set Up allowed Overtime Working at each stage.

If not allowed enter a 0 into the ALLOW OVERTIME Table, if allowed enter the required proportion (Fig. C.4). Set the know demand data for items 1, 2 and 3 into Sheets LINE1, LINE2 and LINE3 at their required delivery dates.

ALLOW OVERTIME 0: no overtime working; 1: full overtime allowed	
Stage	Proportion
1	0
2	0
3	0

Fig. C.4 Loading overtime data into worksheet

DATES	1	2	12	13	14	15	16	NOW 17	18	19	20	21	22	23	24	25	26	27	28	29	30	31
								1														

EXISTING DEMANDSDEMANDS																						
ITEM 1					2						2							3			1	
ITEM 2						1									2							
ITEM 3			3											2								

Fig. C.5 Adding all existing orders to the worksheet

For example here the data for LINE1 contains:

- Time 14 Delivery Demand for 2 items
- Time 20 Delivery Demand for 2 items
- Time 27 Delivery Demand for 3 items
- Time 30 Delivery Demand for 1 item

Adding all existing orders gives Fig. C.5.

Using this data produces plots indicating the feasibility of the existing demand/delivery profile in Fig. C.6.

C.1.5 Using the Program to Check Capacity

We shall now test a proposed new order for feasibility. Add the new order; 2 units of ITEM 2 with a desired delivery date of end of time 26. Figure C.7.

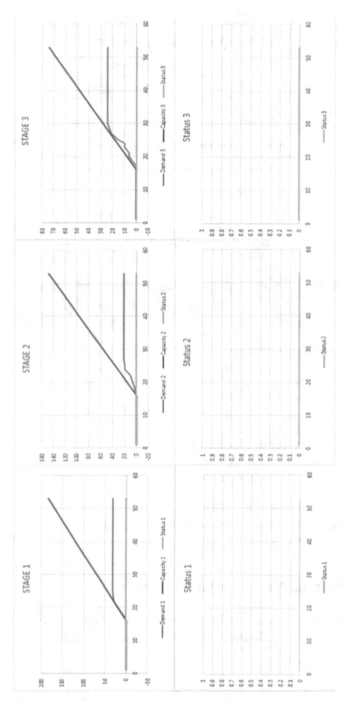

Fig. C.6 Visualising the existing demand/delivery profile

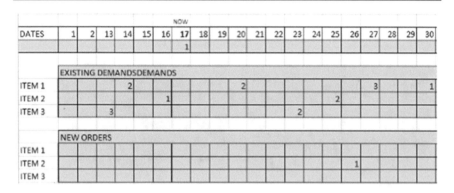

Fig. C.7 Checking the feasibility of a new order

C.1.6 Status Plots

The system produces two plots for each stage, for example for Stage 1 the plots are shown in Fig. C.8. These plots show that without any overtime working this delivery date will not be achievable, other plots fine.

Reschedule this date until a feasible plan/date has been determined, Figs. C.9 and C.10.

C.2 Program 2: Allocating Production Time

Two items of Product 1 are ordered for completion at time 23, progressive allocation.

The information sheet for Product 1 indicates that a time of 23 is possible for the first item (at Stage 3) indicated by the YES in Fig. C.11.

Allocate the first item on Stage 3 at time 23, Into the CELL STAGE3/TIME 23 enter a 1 (Fig. C.12).

Production Stage 3: Place a 1 into Stage 3 (first line) column 32 (Fig. C.13). Proposed delivery date.

Production Stage 2: Place a 1 into column 27 the first feasible free Stage 2 slot (Fig. C.14).

Production Stage 1: Place a 1 into column 25 the first feasible free Stage 2 slot (Fig. C.15).

Update the fixed schedule using these feasible Stage 3, 2 and 1 dates, Fig. C.16.

Repeat the three-stage process to add the second item from this proposed order giving a feasible plan (Fig. C.17).

Update the fixed schedule using these feasible Stage 3, 2 and 1 dates. To give a feasible production plan for this new order for two items of type 1 as in Fig. C.18.

Having assigned the items from this new order summary information is provided on sheets.

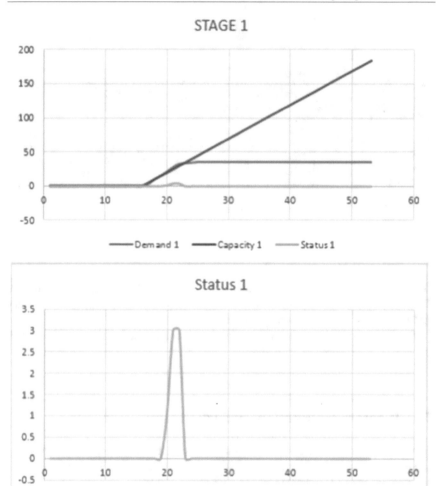

Fig. C.8 Plots to illustrate the status of the schedule

C.2.1   Plan

This gives the stage loading for each time slot, Fig. C.19.

C.2.2   Loading Plan

Indicates the status of each stage at each time, Fig. C.20.

DATES		NOV																			
	1	2	13	14	15	16	17	18	19	20	21	22	23	24	25	26	27	28	29	30	31
							1														
EXISTING DEMANDSDEMANDS																					
ITEM 1			2							2							3			1	
ITEM 2						1								2							
ITEM 3			3										2								
NEW ORDERS																					
ITEM 1																					
ITEM 2																			1		
ITEM 3																					

Fig. C.9 Results of rescheduling

C.3 Assigning Machines to Jobs

The current status of each machine within the workshop is as shown in Fig. C.21 Having received a new order enter (Figs. C.22 and C.23).

1. The expected duration of the job on each machine;
2. Define the possible machines for this job.

Select the most suitable machine using the data displayed in Fig. C.22.

For example, if machine 1 is the selected machine, choose production time (just in time or early as possible) to give the updated schedule (resetting the job duration and machine preference data) as per Fig. C.24. Additional information to assist scheduling is the percentage machine loadings during the period to the due date, Fig. C.25. Alternatively the spreadsheet can be used to select an appropriate delivery date, if for example an order is received at time 1, investigating a finish time at time of 6 leads to an infeasible problem, no possible production slots (Fig. C.26). An acceptable schedule can be obtained by increasing the due date until a feasible slot has been determined, as in Fig. C.27. However if machine 2 is preferred continue increasing the delivery date until an acceptable schedule has been determined (Fig. C.28). This approach giving a set of possible solutions:

1. earliest delivery date 8, using machine 1; or
2. delivery at date 11, using best machine 2.

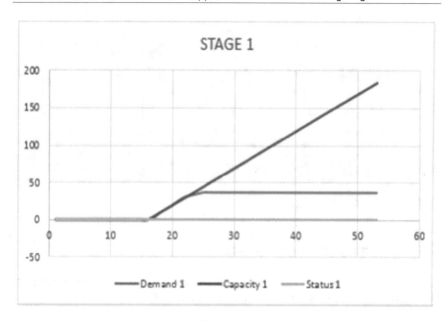

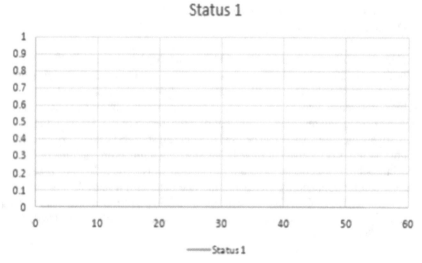

Fig. C.10 Results of rescheduling

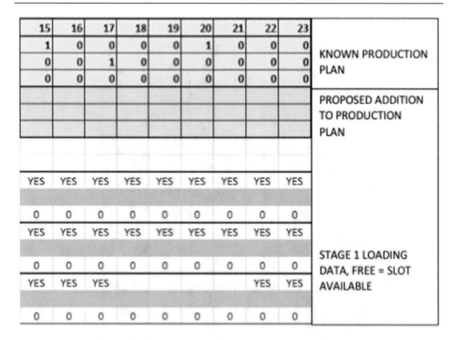

15	16	17	18	19	20	21	22	23	
1	0	0	0	0	1	0	0	0	KNOWN PRODUCTION PLAN
0	0	1	0	0	0	0	0	0	
0	0	0	0	0	0	0	0	0	
									PROPOSED ADDITION TO PRODUCTION PLAN
YES	YES	YES	YES	YES	YES	YES	YES	YES	
0	0	0	0	0	0	0	0	0	
YES	YES	YES	YES	YES	YES	YES	YES	YES	
									STAGE 1 LOADING DATA, FREE = SLOT AVAILABLE
0	0	0	0	0	0	0	0	0	
YES	YES	YES					YES	YES	
0	0	0	0	0	0	0	0	0	

Fig. C.11 Populating the information sheet

15	16	17	18	19	20	21	22	23	
1	0	0	0	0	1	0	0	0	
0	0	1	0	0	0	0	0	0	
0	0	0	0	0	0	0	0	0	
								1	PROPOSED ADDITION TO PRODUCTION PLAN
YES	YES	YES	YES	YES	YES	YES	YES	YES	
0	0	0	0	0	0	0	0	0	
YES	YES	YES	YES	YES	YES				STAGE 1 LOADING DATA, FREE = SLOT AVAILABLE
0	0	0	0	0	0	0	0	0	
YES	YES	YES					YES	YES	
									UPDATED AVAILABLE TIME SLOTS FOR THIS JOB AT STAGE 2.
0	0	0	0	0	0	0	0	0	

Fig. C.12 Allocate the first item on Stage 3 at time 23, Into the CELL STAGE3/TIME 23 enter a 1

	10	11	12	13	14	15	16	17	18	19	20	21	22	23	24	25	26	27	28	29	30	31	32
	0	0	0	0	0	0	0	0	0	0	0	0	0	0	0	0	0	1	0	0	0	0	0
	0	0	0	0	0	0	0	0	0	0	0	0	1	0	0	0	0	0	0	0	0	0	0
	0	0	0	0	0	0	0	0	0	0	1	0	0	0	0	0	0	0	0	0	0	0	0
																							1

	10	11	12	13	14	15	16	17	18	19	20	21	22	23	24	25	26	27	28	29	30	31	32
	YES	YES	YES	YES	YES	YES	YES	YES	YES	YES	YES	YES	YES						YES	YES	YES	YES	YES
	0	0	0	0	0	0	0	0	0	0	0	0	0	0	0	0	0	0	0	0	0	0	0
	YES	YES	YES	YES	YES	YES	YES			YES	YES	YES	YES	YES	YES	YES	YES	YES	YES	YES	YES	YES	YES
	0	0	0	0	0	0	0	0	0	0	0	0	0	0	0	0	0	0	0	0	0	0	0
		YES												YES	YES	YES	YES	YES	YES	YES	YES	YES	YES
	0	0	0	0	0	0	0	0	0	0	0	0	0	0	0	0	0	0	0	0	0	0	0

Fig. C.13 Place a 1 into Stage 3 (first line) column 32

10	11	12	13	14	15	16	17	18	19	20	21	22	23	24	25	26	27	28	29	30	31	32
0	0	0	0	0	0	0	0	0	0	0	0	0	0	0	0	0	1	0	0	0	0	0
0	0	0	0	0	0	0	0	0	0	0	0	1	0	0	0	0	0	0	0	0	0	0
0	0	0	0	0	0	0	0	0	0	1	0	0	0	0	0	0	0	0	0	0	0	0
																	1					1
YES	YES	YES	YES	YES	YES	YES	YES	YES	YES	YES	YES	YES						YES	YES	YES	YES	YES
0	0	0	0	0	0	0	0	0	0	0	0	0	0	0	0	0	0	0	0	0	0	0
YES	YES	YES	YES	YES	YES	YES			YES	YES	YES	YES	YES	YES	YES	YES	YES	0	0	0	0	0
0	YES	0	0	0	0	0	0	0	0	0	0	0	YES	YES	YES			YES	YES	YES	YES	YES
0	0	0	0	0	0	0	0	0	0	0	0	0	0	0	0	0	0	0	0	0	0	0

Fig. C.14 Place a 1 into column 27 the first feasible free Stage 2 slot

10	11	12	13	14	15	16	17	18	19	20	21	22	23	24	25	26	27	28	29	30	31	32
0	0	0	0	0	0	0	0	0	0	0	0	0	0	0	0	0	1	0	0	0	0	0
0	0	0	0	0	0	0	0	0	0	0	0	1	0	0	0	0	0	0	0	0	0	0
0	0	0	0	0	0	0	0	0	0	1	0	0	0	0	0	0	0	0	0	0	0	0
															1		1					1

10	11	12	13	14	15	16	17	18	19	20	21	22	23	24	25	26	27	28	29	30	31	32
YES	YES	YES	YES	YES	YES	YES	YES	YES	YES	YES	YES	YES						YES	YES	YES	YES	YES
0	0	0	0	0	0	0	0	0	0	0	0	0	0	0	0	0	0	0	0	0	0	0
YES	YES	YES	YES	YES	YES	YES			YES				YES	YES	YES	YES	YES					
0	0	0	0	0	0	0	0	0	0	0	0	0	0	0	0	0	0	0	0	0	0	0
													YES	YES	YES			YES	YES	YES	YES	YES
0	0	0	0	0	0	0	0	0	0	0	0	0	0	0	0	0	0	0	0	0	0	0
YES															YES							
0	0	0	0	0	0	0	0	0	0	0	0	0	0	0	0	0	0	0	0	0	0	0

Fig. C.15 Place a 1 into column 25 the first feasible free Stage 2 slot

	10	11	12	13	14	15	16	17	18	19	20	21	22	23	24	25	26	27	28	29	30	31	32
	0	0	0	0	0	0	0	0	0	0	0	0	0	0	0	0	0	1	0	0	0	0	1
	0	0	0	0	0	0	0	0	0	0	0	0	1	0	0	0	0	1	0	0	0	0	0
	0	0	0	0	0	0	0	0	0	0	1	0	0	0	0	1	0	0	0	0	0	0	0

	10	11	12	13	14	15	16	17	18	19	20	21	22	23	24	25	26	27	28	29	30	31	32
	YES	YES	YES	YES	YES	YES	YES	YES	YES	YES	YES	YES	YES						YES	YES	YES	YES	YES
	YES	YES	YES	YES	YES	YES	YES	YES	YES	YES	YES	YES	YES						YES	YES	YES	YES	YES
	0	0	0	0	0	0	0	0	0	0	0	0	0	0	0	0	0	0	0	0	0	0	0
	YES	YES	YES	YES	YES	YES	YES			YES	YES	YES	YES	YES	YES	YES	YES	YES	YES	YES	YES	YES	YES
	0	0	0	0	0	0	0	0	0	0	0	0	0	0	0	0	0	0	0	0	0	0	0
	0	YES	0	0	0	0	0	0		0	0	0	0	0	0	0	0		YES	YES	YES	YES	YES
	0	0	0	0	0	0	0	0	0	0	0	0	0	0	0	0	0	0	0	0	0	0	0

Fig. C.16 Update the fixed schedule using these feasible Stage 3, 2 and 1 dates

10	11	12	13	14	15	16	17	18	19	20	21	22	23	24	25	26	27	28	29	30	31	32
0	0	0	0	0	0	0	0	0	0	0	0	0	0	0	0	0	1	0	0	0	0	1
0	0	0	0	0	0	0	0	0	0	0	0	1	0	0	0	0	1	0	0	0	0	0
0	0	0	0	0	0	0	0	0	0	1	0	0	0	0	1	0	0	0	0	0	0	0
	1																1					1

10	11	12	13	14	15	16	17	18	19	20	21	22	23	24	25	26	27	28	29	30	31	32
YES	YES	YES	YES	YES	YES	YES	YES	YES	YES	YES	YES	YES						YES	YES	YES	YES	YES
YES	YES	YES	YES	YES	YES	YES	YES	YES	YES	YES	YES	YES	YES	YES	YES	YES	YES					
0	0	0	0	0	0	0	0	0	0	0	0	0	0	0	0	0	0	0	0	0	0	0
YES	YES	YES	YES	YES	YES	YES			YES	YES	YES	YES	YES	YES	YES	YES	YES	YES	YES	YES	YES	YES
0	0	0	0	0	0	0	0	0	0	0	0	0	0	0	0	0	0	0	0	0	0	0
0	YES	0	0	0	0	0	0	0	0	0	0	0	0	0	0	0	0	0	0	0	0	0

Fig. C.17 Update the fixed schedule using these feasible Stage 3, 2 and 1 dates

10	11	12	13	14	15	16	17	18	19	20	21	22	23	24	25	26	27	28	29	30	31	32
0	0	0	0	0	0	0	0	0	0	0	0	0	0	0	0	0	1	0	0	0	0	2
0	0	0	0	0	0	0	0	0	0	0	0	1	0	0	0	0	2	0	0	0	0	0
0	1	0	0	0	0	0	0	0	0	1	0	0	0	0	1	0	0	0	0	0	0	0
YES	YES	YES	YES	YES	YES	YES	YES	YES	YES	YES	YES	YES										
0	YES	YES	YES	YES	YES	YES	0	0	YES	YES	YES	YES	YES	YES	YES	0	0	0	YES	YES	YES	YES
YES	YES	YES	YES	YES	YES	YES	0	0	0	0	0	0	0	0	0	0	0	0	0	0	0	0
0	0	0	0	0	0	0	0	0	0	0	0	0	0	0	0	0	0	0	0	0	0	0
0	0	0	0	0	0	0	0	0	0	0	0	0	0	0	0	0	0	YES	YES	YES	YES	YES
0	0	0	0	0	0	0	0	0	0	0	0	0	0	0	0	0	0	0	0	0	0	0

Fig. C.18 Update the fixed schedule using these feasible Stage 3, 2 and 1 dates. To give a feasible production plan for this new order for 2 items of type 1

PRODUCT 1

	11	12	13	14	15	16	17	18	19	20	21	22	23
3	1	0	0	0	1	0	0	0	0	2	0	0	1
2	0	1	0	0	0	1	1	0	0	1	0	0	0
1	0	0	1	0	0	1	0	0	0	0	0	0	0

PRODUCT 2

	11	12	13	14	15	16	17	18	19	20	21	22	23
3	0	0	0	0	0	0	0	0	0	0	0	0	0
2	0	0	0	0	0	0	0	0	0	0	0	1	0
1	1	0	0	0	0	0	0	0	0	1	0	0	0

Fig. C.19 Stage loading for each time slot

SUMMARY Existing PLANS from Product1 and Product2 Pages

		1	2	3	4	5	6	7	8	9	10
	LOADING PLAN ITEM 1										
Stage 3								P1	P1	P1	
Stage 2				P1	P1	P1	P1	P1	P1	P1	P1
Stage 1		P1	P1	P1	P1	P1	P1				P1
	LOADING PLAN ITEM 2										
Stage 3											
Stage 2											
Stage 1								P2	P2	P2	

	TOTAL LOADING										
Stage 3	1							wip	wip	wip	
Stage 2	2			wip	wip	wip	wip	wip	wip	wip	wip
Stage 1	2	wip	wip	wip	wip	wip	wip	wip	wip	wip	wip

STAGE STATUS DETAILS		cap idicates capacity available											
Stage 3	Cap	Cap	Cap	Cap	Cap	Cap	Cap	Cap	Cap	Cap	Cap	Cap	Cap
Stage 2	Cap	Cap	Cap	Cap	Cap	Cap	Cap	Cap	Cap	Cap	Cap	Cap	Cap
Stage 1	Cap	Cap											

Fig. C.20 Status of each stage for each time slot

Time	1	2	3	4	5	6	7	8	9	10	11	12	13	14	15	16	17	18	19	20
Schedule Machine																				
1	1	1	1								1	1		1	1	1				
2	1				1	1	1	1	1					1	1	1				
3	1	1	1									1	1							
4	1				1	1	1	1								1	1	1	1	

Fig. C.21 Current status of machines within the workshop

Job Duration	Job Durations			Machine Restrictions Possible =1			Largest Time Slot Without Rescheduling	
5	1	5		Machine			Machine	Free Slot
DueDate	2	3		1	1		1	6
16	3	6		2	1		2	4
Date	4	5		3			3	
4				4	1		4	8

Fig. C.22 Configuring the expected duration for each job

Time	1	2	3	4	5	6	7	8	9	10	11	12	13	14	15	16	17	18	19	20
										Existing Schedule										
Machine																				
1	1	1	1							1	1			1	1	1				
2	1			1	1	1	1	1						1	1	1				
3	1	1	1									1	1							
4	1			1	1	1	1									1	1	1	1	

Fig. C.23 Configuring the production routing for each job

Job Duration	Job Durations		Machine Restrictions Possible =1		Largest Time Slot Without Rescheduling	
5	1		Machine		Machine	Free Slot
DueDate	2		1		1	
12	3		2		2	
Date	4		3		3	
4			4		4	

Time	1	2	3	4	5	6	7	8	9	10	11	12	13	14	15	16	17	18	19	20
										Existing Schedule										
Machine																				
1	1	1	1	1	1	1	1			1	1			1	1	1				
2	1			1	1	1	1	1						1	1	1				
3	1	1	1									1	1							
4	1			1	1	1	1									1	1	1	1	

Fig. C.24 Updating the schedule

Job Duration	Job Durations			Machine Restrictions Possible =1		Largest Time Slot Without Rescheduling		Percentage Current
5	1	5		Machine		Machine	Free Slot	Loading
DueDate	2	3		1	1	1	6	41
12	3	6		2	1	2	4	50
Date	4	5		3		3		33
4				4	1	4	5	41

Fig. C.25 Entering machine loading during the day to update the schedule

Job Duration	Job Durations			Machine Restrictions Possible =1		Largest Time Slot Without Rescheduling	
5	1	5		Machine		Machine	Free Slot
DueDate	2	3		1	1	1	
6	3	6		2	1	2	
Date	4	5		3		3	
1				4	1	4	

Fig. C.26 Entering additional information

Job Duration	Job Durations			Machine Restrictions Possible =1		Largest Time Slot Without Rescheduling	
5	1	5		Machine		Machine	Free Slot
DueDate	2	3		1	1	1	5
8	3	6		2	1	2	
Date	4	5		3		3	
1				4	1	4	

Fig. C.27 Increasing due date

Job Duration	Job Durations			Machine Restrictions Possible =1		Largest Time Slot Without Rescheduling	
5	1	5		Machine		Machine	Free Slot
DueDate	2	3		1	1	1	6
11	3	6		2	1	2	3
Date	4	5		3		3	
1				4	1	4	

Fig. C.28 Changing preference to machine 2

Scheduling Jobs Through a Workshop

A common problem within a manufacturing firm is to schedule work so that the firms workshop operates both efficiently and effectively, and how improved (computer-based) data can be used to improve the operation of this workshop.

Consequently the aim for the production planner is, given a set of jobs find the best processing order, for these jobs, so that the resultant schedule will be optimal.

However to be able to implement policies, production plans, the planner needs to be able to define the meaning of optimality relevant to the manufacturing firm.

For example is a schedule considered to be 'optimal' if it enables the firm to:

- Finish all jobs as soon as possible (minimise makespan).
- Minimise the total time spent by all the jobs in the workshop.
- Meet delivery times while minimising.

  – total lateness; or
  – total lateness costs; or
  – total earliness; or
  – the maximum job lateness; or
  – minimise finished goods stock holding costs; or
  – minimise the total earliness and lateness costs.

But because these possible objectives can be, are often, in conflict the company will need to determine which, of these, is most relevant to their operations, evidence would suggest that for smaller firms minimising lateness, or lateness costs, will be the most important criterion.

© Springer Nature Switzerland AG 2021
R. Hill and S. Berry, *Guide to Industrial Analytics*, Texts in Computer Science,
https://doi.org/10.1007/978-3-030-79104-9

Table D.1 Number of possible schedules

		Machines, m		
		Flow shop	Job shop 3	Job shop 5
Jobs, n	2	2	8	32
	3	6	216	7776
	4	24	13824	7962624
	5	120	1728000	2.49E+14
	6	720	3.73E+08	1.93E+14

D.1 Scheduling n Jobs Through a Workshop Containing m Machines/Processors/Servers

This section demonstrates how 'optimal' scheduling can be determined calculated using an MS Excel spreadsheet-based program. Note that larger problems, more jobs and/or more production stages, would require the use of a programming language (such as Python) and a heuristic approach to determine a good solution.

Models are presented to represent the scenarios:

- *Permutation Flow Shops*—Each job visits each stage in the same order and the processing order (of the jobs) is the same at each stage.
- *Flow Shop*—Each job visits each stage in the same order but the processing order (of the jobs) is not necessarily the same at each stage.
- *Job Shops*—Each job visits each stage but not necessarily in the same order.
- *Permutation Flow Shops* where the schedule can change at the arrival on a set of new orders.

D.1.1 Problem Size

There are n jobs to be processed through m machines. If all schedules were to be investigated then consider the two extreme cases:

- Permutation Flow Shop, here there are $n!$ schedules to be considered
- Job Shop, here there are $(n!)^m$ schedules to be considered.

Table D.1 illustrates the range of possible schedules. Notice that in a small firm with three production stages and the workshop being organised as a permutation Flow Shop, a common situation is to be able to determine the optimal schedule for six jobs, not a large number of jobs. This would require the consideration of 720 alternative schedules which is difficult to perform manually.

Fig. D.1 Four machines in series

However, the same number of jobs where the workshop is organised as a jobshop would require the consideration of in excess of 300 million alternative schedules which is clearly impossible to conduct manually.

Consequently, for such larger problems the optimal, or good, schedule can be obtained using one of the approaches/techniques based around:

- mathematical programming techniques;
- simulation techniques;
- heuristic approaches.

D.2 Mathematical Programming Techniques

To illustrate this approach to scheduling in a Permutation Flow Shop consider a sample problem consisting of three jobs to be processed through four machines, representing the scale of problem experienced by the majority of small manufacturing firms.

Here, all jobs are processed in the order machine 1 the machine 2 then machine 3 finally machine 4, as shown in Fig. D.1, leading to 24 possible schedules.

D.2.1 Fixing the Schedule Using an Excel Program

To be able to obtain an optimal schedule using the standard solution template requires:

- input data;
- selection of optimality measure.

D.2.2 Data Input: Defining Job Times

Add the job times for each machine/production stage.

For example: Job D1 requires processing times of

- 8 at machine 1;
- 11 at machine 2;
- 7 at machine 3; and
- 8 at machine 4.

Table D.2 Defining the job times for each job, at each machine

	Job, n		
Machine, m	D1	D2	D3
1	8	12	8
2	11	6	8
3	7	13	12
4	8	6	10

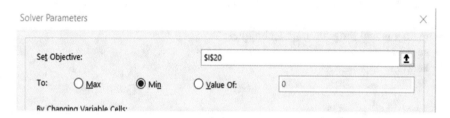

Fig. D.2 Setting the objective in *Solver* in MS Excel

Adding times for the other jobs completes the input data as per Table D.2.

D.2.3   Defining Optimality

Select *Min*imise then give the location of the objective, here the indicated objective has been defined in cell I20 (Fig. D.2).

The objective in cell I20 is to obtain a schedule so that the makespan is minimised. Note that there are four possible objectives available:

1. makespan;
2. total time in system for all jobs;
3. job waiting time in the system.
4. minimise job lateness with respect to agreed delivery dates.

Figure D.3 shows the set-up to minimise wait time, notice that the S columns indicate the start times for each job on each machine/stage and the E columns the end times for each job at each stage. Selecting solve, using mathematical programming, gives the solution to this optimisation problem. This solution is shown in Fig. D.4, the implied schedule is given in the Gantt Chart (see Fig. C.7). This schedule minimises wait time and consequently the time to finish all jobs.

	A	B	C	D	E	F	G	H	I	J	K	L	M	N
1														
2			D1				D2				D3			
3			8				12				8			
4			11				6				8			
5			7				13				12			
6			8				6				10			
7														
8							PRODUCTION SCHEDULE							
9			S		F		S		F		S		F	
10														
11			0		16		0		28		0		8	
12			0		28		0		34		0		16	
13			0		35		0		48		0		28	
14			0		46		0		54		0		38	
15														
16					79				110				52	
17														
18							DURATION and DETAILS							
19			Makespan		Tends		TotTsys		TotalWait					
20			0		0		0		0					
21														
22			0		0		0							
23														

Fig. D.3 Initialising the inputs in the spreadsheet program

D.2.4 Minimise Lateness

Often a company will aim to deliver goods at an agreed date, especially true for a small manufacturing firm delivering directly to their customers, and late deliveries could incur penalty costs. Hence the objective can be to minimise lateness (Fig. D.5).

Delivery deadlines of [60, 50, 40] were imposed onto the data for these three jobs [D1, D2, D3]. This gave an alternative solution where lateness was minimised, see Fig. D.6. Notice that minimising lateness does not imply that the associated makespan will be minimised, compare Fig. D.7 for the schedule defined by this solution and the alternative schedule in Fig. C.10. Here the makespan is 57, all delivered on time, compared with the mathematical programming generated solution with a makespan of 60 all delivered on time. These solutions demonstrating that a program will deliver a solution minimising lateness and so considers both solutions to be equal, although the planner/firm may consider the alternative solution to be better.

Thus, these two models indicate the importance of the criteria, what to optimise, on the form of the solution (Fig. D.8).

D.2.5 Job Shop Scheduling

Here all the jobs may not visit the machines in the same order. Here the input data gives both the processing order and the duration at each machine, see Table D.3. Leading

	D1		D2		D3	
	8		12		8	
	11		6		8	
	7		13		12	
	8		6		10	
Due Dates	60		50		40	
Costs						
Daylate	5		5		5	
Dayearly	1		2		3	

PRODUCTION SCHEDULE

	S	E	S	E	S	E
	8	16	16	28	0	8
	17	28	28	34	8	16
	28	35	35	48	16	28
	38	46	48	54	28	38
Waiting		12		17		0
Lateness/earlyness		-14		4		-2
Lateness		0		4		0
Earlyness		14		0		2

DURATION and DETAILS

Makespan	Tends	TotTsys	TotalWait	Total/Late	Total/Early
54	138	114	29	4	16
				Cost	Cost
54	138	114		20	20

Fig. D.4   Solution

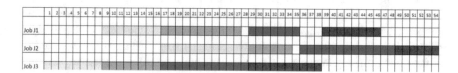

Fig. D.5   Schedule of orders arranged as a Gantt chart

	D1		D2		D3	
	8		12		8	
	11		6		8	
	7		13		12	
	8		6		10	
Due Dates	60		50		40	
Costs						
Daylate	5		5		5	
Dayearly	1		2		3	

PRODUCTION SCHEDULE

	S	E	S	E	S	E
	26	34	8	20	0	8
	34	45	25	31	8	16
	45	52	31	44	16	28
	52	60	44	50	30	40
Waiting	26		13		2	
Lateness/earlyness	0		0		0	
Lateness	0		0		0	
Earlyness	0		0		0	

DURATION and DETAILS

Makespan	Tends	TotTsys	TotalWait	Total/Late	Total/Early
60	150	116	41	0	0
				Cost	Cost
60	150	116		0	0

Fig. D.6 Using delivery deadlines to constrain the model resulting in lateness being minimised

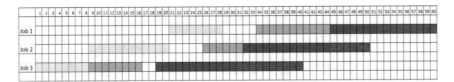

Fig. D.7 Gantt chart minimising lateness

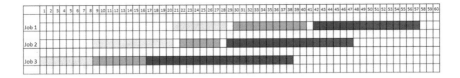

Fig. D.8 Gantt chart showing rearranged solution

Table D.3 Input data for job shop scheduling

Machine, m	Job, n	
	JS1	JS2
1	11	12
2	6	13
3	4	2
4	9	
5	7	8
Due date	70	50

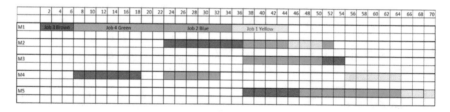

Fig. D.9 Gantt chart for derived job shop schedule

Table D.4 Input data for rescheduling orders in a job shop

Machine, m	Job, n			
	JS1	JS2	JS3	JS4
1	11	12	6	15
2	6	13	11	9
3	4	2	0	11
4	9	6	0	0
5	7	8	0	0
Due date	70	50	60	70

to an optimal schedule for four jobs unlike (by route through the workshop), the resultant schedule, where the aim was to minimise lateness, is displayed in Fig. D.9.

D.3 Rescheduling Jobs

Dynamic scheduling in a Flow Shop can be achieved using this approach used to determine the solution to a job shop scheduling problem.

For example consider a Flow Shop with two jobs in progress, part finished, at the time when two new jobs arrive see Table D.4.

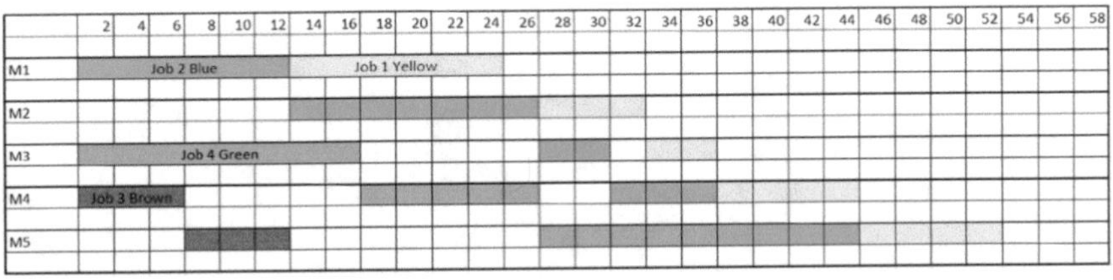

Fig. D.10 Rescheduling the jobs

The resulting schedule is shown in Fig. D.10, notice that here the jobs in progress are scheduled to occur as soon as possible.

Index

A
Apple Numbers, 54
Arrival distribution, 115
Arrival rate, 104, 112
Assumptions, 54

C
Central tendency, 22, 29, 31
Ciw, 103, 115
Clustered, 29
Cold-start, 107
Constant Work In Progress (CONWIP), 76, 79
Continuous, 23, 28
Customer class, 115
Cycle time, 57

D
Data quality, 23
Demand pattern, 101
Descriptive statistics, 22
Discrete, 23
Dispersed, 31
Dispersion, 22, 30, 31
Distribution, 94
Dynamic system, 101

E
Embedded systems, 24
Experiment, 101
Experimentation, 97
Exponential distribution, 103

F
Ford Motor Company, 65
Forecast, 99
Forecasting, 56
Frequency, 22

G
Generic planning and control, 75
Google Sheets, 54
Greedy, 69

I
Industrial process, 98, 101
Inferential statistics, 22
Interquartile Range (IQR), 32

J
Joinery manufacturer, 102
Just-in-time, 66

K
Kanban, 65

L
Lead time, 57, 119
Lean manufacturing, 96
Lean production, 71
Little's law, 56

M
Make-to-order, 102

© Springer Nature Switzerland AG 2021
R. Hill and S. Berry, *Guide to Industrial Analytics*, Texts in Computer Science,
https://doi.org/10.1007/978-3-030-79104-9

Manufacturing cell, 102
Manufacturing plant, 113
Manufacturing system, 111
Mean, 29
Median, 29
Microsoft Excel, 54
Mode, 29
Model, 88, 92
Modelling, 98, 101
Model T Ford, 65
Modular approach, 102
MRP, 76

N
Nominal, 25

O
Operational research, 53
Operations management, 54
Opinion survey, 28
Optimised Production Technology (OPT), 76
Ordered data, 32
Ordinal data, 28
Outlier, 30

P
Percentage, 22
Process improvements, 55
Production processes, 101
Production routing, 105
Product routing, 112
Proportion, 22

Q
Qualitative, 22
Qualitative and quantitative data, 22
Quantitative, 22
Queue, 89
Queue capacity, 106
Queueing, 98, 101
Queueing approach, 103
Queueing model, 98, 103
Queueing network, 102

R
Ratio, 30

Regression modelling, 71
Resource utilisation, 111, 112
Rouge River Ford plant, 65
Routing, 105
Routing matrix, 106

S
Sales order processing, 102
Scale, 25
Scientific management, 65
Seed, 97
Servers, 93, 94
Service distribution, 94, 105
Service rate, 103
Service time, 96
Simulation, 80, 92, 93
Skewness, 22
Small data, 24
Spread of data, 31
Stakeholder, 54
Strategic, 99
Summary statistics, 22
System, 102

T
Takt time, 57
Taylor, 65
Theory of Constraints (TOC), 76
Throughput, 56
Time and motion, 65
Toyota, 65

U
Utilisation, 89, 95, 96, 107
Utilisation of resource, 113

V
Variance, 32
Visibility, 102

W
Wait time, 96
Waste, 96
Work In Progress (WIP), 55, 107, 116
Workstation, 103

Printed in the United States
by Baker & Taylor Publisher Services